T0193163

Game Theory

A Classical Introduction, Mathematical Games, and the Tournament

Synthesis Lectures on Games and Computational Intelligence

Editor
Daniel Ashlock, *University of Guelph*

Game Theory: A Classical Introduction, Mathematical Games, and the Tournament
Andrew McEachern
2017

Game Theory: A Classical Introduction, Mathematical Games, and the Tournament
Andrew McEachern

ISBN: 978-3-031-00990-7   paperback
ISBN: 978-3-031-02118-3   ebook

DOI 10.1007/978-3-031-02118-3

A Publication in the Springer series
SYNTHESIS LECTURES ON GAMES AND COMPUTATIONAL INTELLIGENCE

Lecture #1
Series Editor: Daniel Ashlock, *University of Guelph*
Series ISSN
ISSN pending.

Game Theory

A Classical Introduction, Mathematical Games, and the Tournament

Andrew McEachern
Queen's University

SYNTHESIS LECTURES ON GAMES AND COMPUTATIONAL INTELLIGENCE #1

ABSTRACT

This book is a formalization of collected notes from an introductory game theory course taught at Queen's University. The course introduced traditional game theory and its formal analysis, but also moved to more modern approaches to game theory, providing a broad introduction to the current state of the discipline. Classical games, like the Prisoner's Dilemma and the Lady and the Tiger, are joined by a procedure for transforming mathematical games into card games. Included is an introduction and brief investigation into mathematical games, including combinatorial games such as Nim. The text examines techniques for creating tournaments, of the sort used in sports, and demonstrates how to obtain tournaments that are as fair as possible with regards to playing on courts. The tournaments are tested as in-class learning events, providing a novel curriculum item. Example tournaments are provided at the end of the book for instructors interested in running a tournament in their own classroom. The book is appropriate as a text or companion text for a one-semester course introducing the theory of games or for students who wish to get a sense of the scope and techniques of the field.

KEYWORDS

classical game theory, combinatorial games, tournaments, deck-based games, tournament design, graph games

Contents

Preface xi

Acknowledgments xiii

1   Introduction: The Prisoner's Dilemma and Finite State Automata 1
   1.1   Introduction 1
       1.1.1 Helpful Sources of Information 2
   1.2   The Prisoner's Dilemma 2
   1.3   Finite State Automata 5
   1.4   Exercises 8

2   Games in Extensive Form with Complete Information and Backward
Induction **9**
   2.1   Introduction 9
   2.2   The Lady and the Tiger Game 10
   2.3   Games in Extensive Form with Complete Information 13
   2.4   Backward Induction 14
   2.5   The Ultimatum Game 15
       2.5.1 The Enhanced Ultimatum Game 16
       2.5.2 What Backward Induction Says 16
       2.5.3 Backward Induction is Wrong About This One 17
   2.6   The Boat Crash Game 20
   2.7   Continuous Games 23
       2.7.1 Two Models of the Stackelberg Duopology 23
   2.8   The Failings of Backward Induction 27
   2.9   Exercises 27

3   Games in Normal Form and the Nash Equilibrium **31**
   3.1   Introduction and Definitions 31
   3.2   The Stag Hunt 31
   3.3   Dominated Strategies 33

        3.3.1 Iterated Elimination of Dominated Strategies 33

  3.4 The Nacho Game 33

        3.4.1 The Nacho Game with K Players 34

  3.5 Nash Equilibria 35

        3.5.1 Finding the NE by IEDS 36

        3.5.2 IEDS process 36

  3.6 The Vaccination Game 37

        3.6.1 The N-player Vaccination Game 39

  3.7 Exercises 39

4   **Mixed Strategy Nash Equilibria and Two-Player Zero-Sum Games** **43**

        4.0.1 The Fundamental Theorem of Nash Equilibria 44

  4.1 An Example with a 3-by-3 Payoff Matrix 45

  4.2 Two-Player Zero Sum Games 47

        4.2.1 The Game of Odds and Evens 47

  4.3 Domination of Two-Player Zero Sum Games 48

        4.3.1 Saddle Points 48

        4.3.2 Solving Two-by-Two Games 49

  4.4 Goofspiel 50

  4.5 Exercises 50

5   **Mathematical Games** **53**

  5.1 Introduction 53

        5.1.1 The Subtraction Game 53

  5.2 Nim 55

        5.2.1 Moore's Nim 56

  5.3 Sprouts 57

  5.4 The Graph Domination Game 61

  5.5 Deck-based Games 63

        5.5.1 Deck-based Prisoner's Dilemma 64

        5.5.2 Deck-based Rock-Paper-Scissors(-Lizard-Spock) 66

        5.5.3 Deck-based Divide the Dollar 68

        5.5.4 FLUXX-like Game Mechanics 69

        5.5.5 A Note on Adding New Mechanics to Mathematical Games 72

  5.6 Exercises 72

6   Tournaments and Their Design **75**
  6.1   Introduction 75
          6.1.1   Some Types of Tournaments 75
  6.2   Round Robin Scheduling 77
  6.3   Round Robin Scheduling with Courts 78
          6.3.1   Balanced Tournament Designs 78
          6.3.2   Court-balanced Tournament Designs 80
  6.4   Cyclic Permutation Fractal Tournament Design 82
          6.4.1   Case 1: $n = 2^k, k \in \mathbb{Z}^+$ 83
          6.4.2   A Recursive Generation of the Minimum Number of Rounds 88
  6.5   Exercises 88

7   Afterword **91**
  7.1   Conclusion and Future Directions 91

A   Example Tournaments **93**
  A.1   Example Tournaments 93
          A.1.1   The Enhanced Ultimatum Game Tournament 93
          A.1.2   The Vaccination Game Tournament 95
          A.1.3   A Different Kind of IPD Tournament 97
  A.2   Some Things to Consider Before Running a Tournament in a Classroom ... 98

Bibliography **101**

Author's Biography **103**

Preface

This book is nothing like a typical introductory game theory book. While it offers some of the same material you would find there, it takes a very different approach in an attempt at being accessible to students and readers from a variety of backgrounds. It is intended for students in the arts or sciences who are interested in a mathematics course that exposes them to the interesting parts of game theory without being strongly focused on rigor. This book does not require a strong mathematics background to follow, but it is meant to be a way to learn the concepts in addition to attending class. Due to the brief nature of this book, there will be many parts that seem incomplete, or rather, like they could be expanded on a great deal. Attending class is the best way to fill those gaps. This book also discusses many topics that have never been considered by game theory researchers. There may be some people who disagree with my inclusion of these topics, but to them I say that this book is about taking a step in a new direction to bring some life and innovation to the game theory field. The chapters on mathematical games have been included because they encourage mathematical thinking while being accessible, and because playing actual games is fun. I've always thought classical game theory has suffered from not including the playing of games, but rather treating games as mathematical objects to be analyzed, categorized, and then placed on the shelf. Finally, the mathematics of tournament design is a subject that has been buried deep in the combinatorics literature. It is time to bring it to the light, applying it in interesting ways to create tournaments, and then conducting those tournaments in all of their optimized glory. Competition is such a driving force in so many aspects of our lives, including playing simple games, that it seems wrong somehow that we do not apply our mathematical techniques to improving it. As we apply these mathematical techniques to sharpening competition, competition sharpens our mathematical thinking, creating a cycle that could bring benefits of which we are not yet aware. This book is the first step in the direction of using games to broaden our intellect and drive us to higher performance. Much of this book features topics from active research, which is another reason why it diverges rather strongly from a typical game theory text.

Andrew McEachern
May 2017

Acknowledgments

I'd like to thank Morgan & Claypool for giving me the opportunity to collect my thoughts and put them down on paper and Daniel Ashlock for his immeasurable support and insights in creating this work.

Andrew McEachern
May 2017

Introduction: The Prisoner's Dilemma and Finite State Automata

1.1   INTRODUCTION

Game theory is a multidisciplinary field that has been pursued independently in economics, mathematics, biology, psychology, and diverse disciplines within computer science (artificial intelligence, theory of computer science, video game development, and human computer interface, to name just a few). Beginning with the work of John Von Neumann and Oskar Morgenstern, *Theory of Games and Economic Behavior* [8], game theory has found its way into several subjects. Why should this be the case, if it's initial focus was on problems in economics?

It may help if we define what we mean when we say **game**. After scouring both texts and the Internet, I found many similar definitions of game theory, but none which captured the essence of the *game* part of game theory. I offer my own definition; a **game** is any situation involving more than one individual, each of which can make more than one action, such that the outcome to each individual, called the *payoff*, is influenced by their own action, and the choice of action of at least one other individual. One way to think about game theory as a field is that it is a collection of theories and techniques that help us think about how to analyze a game. Taking a moment to reflect on this definition, it should be clear why game theory has seen application in so many diverse fields.

The purpose of treating situations like games also changes from research area to research area. In economics, biology, and psychology, the purpose of treating situations involving agents with possibly differing goals is to predict human or animal behavior. By solving the game, and what that means will be defined later in this chapter, a researcher may believe they have a good idea of how an organism (animals, people, or companies) will behave when confronted with a situation similar to their game. For a mathematician, the solving of a game usually refers to an impartial combinatorial game, as you will see in Chapter 5, or proving a theorem about a class of games that gives an overarching idea about what properties they may have. For a computer scientist interested in artificial intelligence, games are an excellent way to study evolution of competitive or cooperative AI under very specific conditions.

1.1.1   HELPFUL SOURCES OF INFORMATION

Depending on the prerequisites of this course, it is possible that you have not been exposed to much of the background material necessary to succeed in fully understanding the main ideas in this course. Here are a list of websites that either contain the supporting material, or will at least point you in the right direction.

- If you are not familiar with basic probability, or would like to brush up, go to `https://www.khanacademy.org/math/probability/probability-geometry /probability-basics/v/basic-probability` is an excellent website.

- Following on basic probability, if you are totally unfamiliar with decision theory, go to `http://people.kth.se/~soh/decisiontheory.pdf` to get a basic introduction.

- To get a handle on the Gaussian, or Normal, distribution, go to `http://www.statisti cshowto.com/probability-and-statistics/normal-distributions`

- If it has been a while since you've seen calculus, mostly with regards to derivatives and partial derivatives, `http://tutorial.math.lamar.edu/` may be the best free collection of calculus notes I have seen.

- If you are unfamiliar with the concept of noise with regards to signals, `https://en.wik ipedia.org/wiki/Noise_(signal_processing)` is a good place to start.

- For help with basic combinatorics and introductory graph theory concepts, you should take a look at Chapters 1 and 4 of `https://www.whitman.edu/mathematics/cgt_on line/cgt.pdf`.

1.2   THE PRISONER'S DILEMMA

We begin our study of classical game theory with a discussion of the Prisoner's Dilemma, since it is possibly the most studied game across several areas in the field of game theory. This is likely due to the fact that it is fairly simple, and it can model a broad range of situations. To list all of the situations the Prisoner's Dilemma has been used to model is not feasible, a search in the field in which you are interested can easily give you an idea of how often this game has been used. This brings up an interesting point to consider, which we should keep in mind for the rest of the time we are studying game theory: George Box, a statistician, wrote that "essentially, all models are wrong, but some are useful." Any time we are modeling a situation with a game, we have to make some simplifying assumptions. When we do this, we lose information about that situation. There are times when games have been used to model a situation involving people, and those games have entirely failed to predict how humans will behave. This is not to say that we should never use games as models, for they are useful, but we must be aware of their limitations.

The original Prisoner's Dilemma revolves around the following story. Two criminals accused of burglary are being held by the authorities. The authorities have enough evidence to convict either criminal of criminal trespass (a minor crime), but do not have enough evidence to convict them of burglary, which carries a much heavier penalty. The authorities separate the accomplices in different rooms and make the following offer to both criminals. The authorities will drop the criminal trespass charge and give immunity from any self-incriminating evidence given, as long as the suspect gives up his or her partner. Four possible outcomes result.

1. Both suspects keep quiet, serve a short sentence for criminal trespassing, and divide the money.

2. One suspect testifies against the other, but his or her partner says nothing. The testifier goes free and keeps all of the money, while the one who kept quiet serves a much longer sentence.

3. The vice versa of possibility 2.

4. Both suspects give the other one up in an attempt to save themselves. They both receive moderate sentences and, assuming they don't kill each other in prison, when they get out they still have a chance to split the money.

It is convenient to quantify the outcomes as payoffs to either player. Since there are two choices for each criminal, we call the first choice of not speaking with the authorities *cooperate*, or **C**. There is a mismatch of terminology here, since we would think cooperate would involve cooperating with the authorities, but instead think of it as cooperating with the other person involved in the game, namely, your partner in crime. The other choice, which is testify against your partner, is called *defect*, or **D**. We can create matrix to represent these payoffs with respect to the choices of both criminals.

$$
\begin{array}{ccc}
& \text{Criminal 2} & \\
& \text{C} \qquad \text{D} & \\
\text{Criminal 1} \quad \text{C} & (3,3) \quad (0,5) & \\
\text{D} & (5,0) \quad (1,1) &
\end{array}
$$

Figure 1.1: The payoff matrix for the Prisoner's Dilemma.

From Figure 1.1, we have quantified the situation as follows: If both criminals keep quiet, or cooperate, they both receive a payoff of 3. If Criminal 1 defects and Criminal 2 cooperates, then Criminal 1 receives a payoff of 5 and Criminal 2 receives a payoff of 0. If the choices are reversed, so are the payoffs. If both attempt to stab each other in the back, defect, the payoff to each is 1. From now on we will call the decision makers in the Prisoner's Dilemma *players*. Any game with two players that has the following payoff matrix as displayed in Figure 1.2 is called *a* Prisoner's Dilemma. C is the payoff when both players cooperate, S is called the sucker payoff, T is called the temptation payoff, and D is the mutual defect payoff.

Player 2

		C	D
Player 1	C	(C,C)	(S,T)
	D	(T,S)	(D,D)

Figure 1.2: The generalized payoff matrix for the Prisoner's Dilemma.

C, S, T, and *D* must satisfy the following inequalities:

$$S \leq D \leq C \leq T \text{ and } (S + T) \leq 2C.$$

The rational thing to do is defect, even though both players will do fairly well if they both cooperate. Once we find out what both players should do, assuming they are operating rationally, we can informally call that the *solution* to the game. Why should this be the case? I leave it to you, the reader to figure it out. More interestingly, if we are to imagine the Prisoner's Dilemma as a model for transactions of many kinds, most people cooperate rather than defect, even without any personal negative consequences that would result from their choice.

- Game theory was invented by John von Neumann and Oskar Morgenstern.[1] At the time John von Neumann was inventing it, he was also advising the U.S. government on national security issues. He concluded that a nuclear strike was the only rational decision [9]. Thankfully for the world in this one case, politicians are not rational.

The Iterated Prisoner's Dilemma

There was a flaw in John von Neumann's reasoning about the Prisoner's Dilemma as a model for interaction between governments, and with people in general.

- Interactions between governments and between people are rarely one shot deals.

Consider the following scenario. In the Prisoner's Dilemma, imagine that Players 1 and 2 defect in the first round, both getting a payoff of 1. Now they face each other once again, and they remember with whom they just played, and the move of their opponent. In the next round, they both have no incentive to cooperate, so they both defect again, and again receive a payoff of 1. Their average score over the two rounds would be 1. Indeed, over any number of rounds the average payoff to the two players would be 1. Now consider the average payoff to the two players if they both cooperated with one another each round. It should be easy to see that the average payoff would be 3. A payoff of 3 isn't as good as a payoff of 5, but it is still better than the result of always defecting. It is better on average to cooperate any time a player will encounter another player more than once, if you know your opponent is likely to cooperate with you.

[1]Von Neumann and Morgenstern [8].

A game which is played more than once between the same two players is called an *iterated game*, and the iterated version of the Prisoner's Dilemma is entirely unlike the one-shot version. This tends to be the case with almost every game in game theory. The Iterated Prisoner's Dilemma comes from *The Evolution of Cooperation* by Robert Axelrod, an excellent book in all regards. He also wrote two papers, [1, 2].

1.3   FINITE STATE AUTOMATA

The Iterated Prisoner's Dilemma (IPD) allows for many more strategies than the one-shot version; we need a systematic way of encoding those strategies. Ultimately, the goal is to have tournaments where students submit strategies on how to play the IPD and finite state automata are a natural encoding to use. They allow a rich variety of strategies and are not difficult to understand and specify. A finite state automaton (FSA) is constituted by the following: an input alphabet, an output alphabet, a finite number of states, a transition function, a response function, an initial state, and an initial response, [6, 10]. The FSA uses states as a form of memory, the transition function defines how the FSA moves from one state to another based on some input, and the response function defines what the FSA will output as it transitions from state to state. The standard encoding of a FSA is a look-up table, but that is only relevant to the would be computer scientists who want to write their own versions. For our purposes, we will consider FSA that can only be in one state at any given time. An example of an everyday FSA will help clarify things.

- An elevator inside of an apartment or office building is a FSA.

- The possible states are the floors to which the elevator can go.

- The input alphabet are the floors of the building, as buttons. The output alphabet is the floors of the building.

- The transition function consists of a button being pushed, presumably by a person wanting to go to a certain floor.

- The response function is the elevator moving to that floor.

- The initial state would be the lobby, and the initial response would also be the lobby.

There are two ways to collect and organize the information that constitutes a FSA. The first is a table, and the second is a diagram. We'll restrict our attention to a two-floor elevator, otherwise both the table and the diagram begin to grow dramatically. The table and diagram are shown in Figure 1.3.

The diagram is created using the *Mealy* architecture. Notice that the arrows define the transition function, and the characters attached as labels define the response function. For a

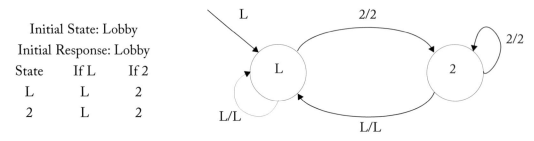

Initial State: Lobby

Initial Response: Lobby

State	If L	If 2
L	L	2
2	L	2

Figure 1.3: The table and diagram of a two-floor elevator as a FSA.

given label A/B, A is the input the FSA receives, and the B is the response that it produces from the output alphabet.

Without exaggeration, there are thousands of applications for which FSAs are appropriate. FSAs have been used in string prediction and classification, as agents in Iterated Prisoner's Dilemma experiments, and in experiments with agent evolution. FSAs have also been used in creating industrial process controllers, and developing intelligent video game opponents. We will restrict our attention to FSAs as agents that play the Iterated Prisoner's Dilemma.

FINITE STATE AUTOMATA AS AGENTS THAT PLAY THE ITERATED PRISONER'S DILEMMA

Before we discuss the material indicated by the section title, we'll first develop the idea of a *strategy*. The word strategy has more than one definition, and some of them have militaristic colorings, but for our purposes we will restrict our definition of strategy in the following manner: a strategy consists of a defined output as a response to a defined input based on a given set of conditions. A FSA is a strategy, by this definition. Often, in tournament settings, strategies will involve differing responses to the same input, depending on the current conditions. A strategy for the IPD made famous by Robert Axelrod's tournaments is called *Tit-for-Tat* (TFT). The strategy of TFT consists of the following: begin the game by cooperating, then do what your opponent did in the previous round. If we capture TFT in table form, it looks like the Figure 1.4.

Tit-for-Tat

Initial Response: C

Initial State: 1

State	If C	If D
1	C→1	D→1

Figure 1.4: Tit-for-Tat.

Axelrod held two tournaments for the IPD, and both times TFT came in first, among strategies submitted by people from a variety of fields. Axelrod reported that the success of TFT was based on four conditions. The first condition is that TFT is a *nice* strategy; it never defects first against an opponent. The second condition is that TFT is *provocable*, meaning it will defend itself if attacked. The third condition is that TFT is *forgiving*, meaning once its opponent starts playing cooperating, TFT will once again resume cooperating. Lastly, TFT is *simple*, meaning other strategies can adapt to it. Some other strategies are listed below.

Always Cooperate

Initial Response: C

Initial State: 1

State	If C	If D
1	C→1	C→1

Tit-for-Two-Tats

Initial Response: C

Initial State: 1

State	If C	If D
1	C→1	C→2
2	C→1	D→2

Always Defect

Initial Response: D

Initial State: 1

State	If C	If D
1	D→1	D→1

Psycho

Initial Response: D

Initial State: 1

State	If C	If D
1	D→1	C→1

Figure 1.5: Various Strategies for the IPD.

Tit-for-Two-Tats is like Tit-for-Tat, but more forgiving, and Psycho does the opposite of whatever its opponent did last time.

A NOTE ON USING FSAS AS AGENTS FOR THE IPD

While this book focuses on using FSAs, there is no reason to think that these are the only kind of representation a person could use. Neural nets, evolvable programs such as ISAc lists, simple and complex programs have all been used as agents in the IPD. A great deal of work has been done with FSAs in the field of evolutionary computation [25]. Evolutionary computation is a field that combines Darwin's theory of biological evolution, computer science, and mathematics, and inspiration from many different fields of study, mainly biology. FSAs are a convenient representation, and can implicitly encode memory into a strategy, rather than having to do it explicitly. Neural nets are also capable of implicit memory encoding, but the author's experiences are mainly to do with FSAs.

1.4 EXERCISES

1. Show that when Psycho plays TFT, TFT can do no better than tie with Psycho. Also show that if those two strategies are playing IPD with a random number of rounds, on average the expected difference in payoffs is positive for Psycho and negative for TFT.

2. Consider the situation in which an Iterated Prisoner's Dilemma tournament involves two TFT agents and two Psycho agents. If we take the average score of all the games played and let that be an agent's tournament score, which strategy does better in this situation? If I change the number of Psycho agents by plus or minus one, how does that affect every agent's average payoff?

3. Create a FSA diagram for the following strategy: start by defecting. If opponent defects three times in a row, cooperate on the next round. If opponent also cooperates, keep cooperating unless opponent defects again. If opponent defects, reset to defecting three times in a row. This strategy has a name: Fortress 3. Find a strategy that defeats Fortress 3 in tournament play.

4. Fortress X describes a family of strategies in which an agent will defect X times, and then cooperate if their opponent cooperates after those X initial defects. As of yet, there has been little investigation in how Fortress X fares against Fortress Y. Compare Fortress 3 with Fortress 4. When these strategies face one another in a tournament, which one does better on average during a match if they play between 145 and 155 rounds, if the number of rounds is chosen uniformly at random?

5. Create or obtain software that runs an Iterated Prisoner's Dilemma tournament. Using 150 rounds plus or minus 5 rounds uniformly at random, enter Fortress 3, TFT, Psycho, and Random into the tournament and examine the results. Run the tournament a few times to get a feel for the distribution of outcomes. Now add two of each strategy and examine the results again. Is there a difference? Which strategy wins in the case where two of each are entered into the IPD?

6. Consider the tournament of Exercise 5. Enter two TFT strategies, and 1 of each strategy listed above. What are the results of the tournament? Do both TFT strategies come out ahead? Give an analysis of the results, including why TFT loses. Now consider the case where there are only 2 TFT strategies, 1 F3, and 1 Random strategy. Prove that under most conditions two TFT strategies will outperform the other two strategies in that tournament. Find the conditions that cause the two TFT strategies to lose to F3, and find the probability those conditions will occur.

CHAPTER 2

Games in Extensive Form with Complete Information and Backward Induction

2.1 INTRODUCTION

There is a story called the *Lady and The Tiger*, written by Frank Stockton in 1882. It's a charming tale and well worth reading, but here is the short version.

A semi-mad barbarian king builds a coliseum in the tradition of his Latin neighbors. The coliseum holds two functions. The first is for gladiatorial games against man and beast, to delight his subjects and appease the king's violent whims. The second use is the far more interesting one. Most crimes in the king's domains were handled by appointed magistrates, and rarely ever reached the king's ears. But crimes of an especially curious—or heinous—nature were brought to the king's attention. In a fit of madness, or perhaps genius, the king devised a way to determine the guilt of the accused. He built two doors into the base of his coliseum, opening inward toward the sands. There was nothing overly special about these doors, excepting the fact that they were identical in every way except position. One was placed to the left of the king's podium, and the other to the right of the king's podium. Behind one door was a ravening tiger, who would presumably eat the accused if he or she chose that door. Behind the other door awaited someone of equal station or wealth, whom the accused would have to marry if they chose that door. By the king's decree, if a person was already married, that previous marriage would be voided by the new one. In this way the crimes that interested the king were solved, and the people thought it was just. After all, didn't the accused pick their own fate?

The king had a daughter, at least as intelligent and ferocious as her father. She was to be wed to one of the princes from a neighboring kingdom, but she fell in love with a lowborn adventurer, and they had an illicit affair. As these things tend to go, the court found out about it and the adventurer was brought before the king. After all, his crime was most assuredly in the purview of the king's interest, regardless of how his daughter felt about the man. The king sent him into the coliseum, and bid him choose his fate. Even the king did not know behind which door lay the tiger, or the adventurer's prospective bride, since the chance delighted him as much as anything else. However, the daughter found out just before the trial was to commence, and as she sat on the king's left, her and her beloved locked eyes. The princess knew that behind one of the doors was the biggest, angriest tiger her father could find, a noted man-eater. Behind the other one was her rival in the politics of the court, a woman she hated with a deep passion. More importantly, the princess knew the location of the tiger and the other lady.

As they maintained their gaze, the princess ever so subtly raised her right hand and put it down again before anyone else could notice, and the king bade the adventurer to choose his fate. Stockton ends the story before revealing the choice and fate of the adventurer, thus leading to the question, "Well, what's it going to be? The lady, or the tiger?"

Apart from being entertaining, this tale has some interesting ideas that we can capture mathematically.

2.2 THE LADY AND THE TIGER GAME

First, we note that there are two players in this game: the princess and the adventurer. All other characters have already made their decisions in the story. Let's assume that the princess has not raised her right hand yet, and she is still deciding on what to do.

- She has two choices: raise her right hand, or raise her left hand.

- The adventurer has two choices as well: choose the left door, or choose the right door.

Since the princess goes first, we can create a diagram called a game-decision tree to help us visualize the possibilities. Consider Figure 2.1.

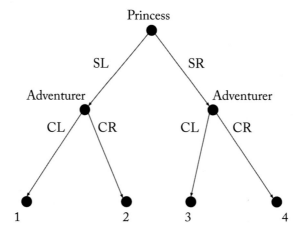

Figure 2.1: The game decision tree for the lady and the adventurer. SL and SR are for the princess' two choices, show left hand or right hand, respectively. CL and CR are for the adventurer's two choices choose the left door and choose the right door.

The vertices, or nodes, of the tree are labeled with the players of the game. At a given vertex, if that vertex is labeled with a certain player, that means the player gets to make a choice.

There may be more than two choices available to a given player. In the case of Figure 2.1, the princess gets to make the first decision, whether to show her left hand (SL) or her right hand (SR). After she makes her choice, the adventurer makes his decision next, and that is either choose the left door (CL) or choose the right door (CR). The *outcomes* of the game are given by the leaves of the tree, and for space consideration they are labeled 1, 2, 3, and 4. The outcomes describe the choices that the players make, and more importantly, the *payoffs*. Payoffs will be formally defined later, but for now think of it as what a player will walk away with when the game is ended.

- Outcome 1 is the result of the princess showing her left hand, and the adventurer choosing the left door.

- Outcome 2 is the result of the princess showing her left hand, and the adventurer choosing the right door.

- Outcome 3 is the result of the princess showing her right hand, and the adventurer choosing the left door.

- Outcome 4 is the result of the princess showing her right hand, and the adventurer choosing the right door.

What should the princess and adventurer do? Our current set-up can't help us because we don't have enough information. We need to make some further assumptions about the scenario.

- Assume the tiger is behind the right door, and the lady is behind the left.

- Assume the adventurer thinks the princess is indicating which door the tiger is behind when she raises her hand.

We must now determine the payoffs for each outcome. In Outcome 1, the princess indicates the left door, and the adventurer chooses that one. The adventurer thought the tiger was behind the right door in that scenario, and so did not trust the princess. We can probably conclude their relationship is dead, and the princess' payoff will likely be that she is sad that she lost him to the other woman, we will denote how sad she is by the variable $-S$. The adventurer's payoff was that he got married, denoted by m, and we'll refrain from describing his emotional state at that. We can list the payoff of Outcome 1 as the ordered pair $(-S, m)$, since the princess is the player that moves first in the game, and the adventurer moves second.

Outcome 2 has the princess indicating the left, and the adventurer choosing the right. The payoffs from that scenario are $-S$ for the princess again, since the adventurer dies and she loses him permanently, and $-D$ for the adventurer, since he is dead. Outcome 2 has the payoffs $(-S, -D)$.

Outcome 3 is the adventurer getting married and believing the princess when she indicates the tiger is behind the right door. The payoff for the princess is $-s$, and I use lowercase s to demonstrate that while she'll be sad the adventurer is getting married to her rival, he trusted her and didn't die. I'm also a bit of an optimist and like to think that they will continue their affair more carefully and continue being in love. So her sadness is not the same level of sadness as in outcome 1. We can say that $S > s$, in this case. The adventurer gets married, so his payoff is still m. The payoffs for Outcome 3 are $(-s, m)$.

Finally, Outcome 4 is the princess indicating the right door, and the adventurer choosing the right door, and being eaten. The payoffs are $(-S, -D)$.

We can quickly capture all of this information in the game decision tree given in Figure 2.2.

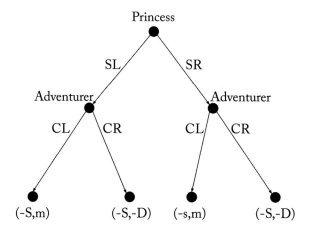

Figure 2.2: The game decision tree for the lady and the adventurer with payoffs.

What should the princess and the adventurer do?

We can use the tree we've developed to help us find the answer to that question. First, we have to use the idea that the princess chooses first, and the adventurer chooses second. So the princess is going to use the fact that she knows all of the outcomes based on whatever decision the adventurer makes, and figure out which one he will choose. Whether she signals right or left, if the adventurer chooses the left door his payoff is better than if he chooses the right door, assuming he finds the prospect of marriage better than death ($m > -D$). If that wasn't the case, our analysis would be a little different, and perhaps a good deal sadder. The princess now knows for certain that the adventurer will choose the left door. We can record the choice of the adventurer in both cases, choose left (CL), and delete the bottom part of the tree as indicated in Figure 2.3.

Now that we've dealt with the adventurer's decisions, we have to examine the princess' choices. If she indicates the left door (SL), her payoff will be $-S$. If she indicates the right door, her payoff will be $-s$. Since $-S < -s$, we determine that she will choose to indicate the right door (CR). This leaves her with a payoff of $-s$, and the adventurer will receive a payoff of m. The solution to our game based on our earlier assumptions is the most optimistic scenario—in

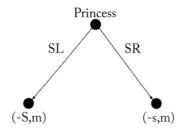

Figure 2.3: The game decision tree for the lady and the adventurer with payoffs.

the author's opinion—and so there is a somewhat happy ending to the story. If the payoff values were different, it is possible that a different outcome would be the solution based on backward induction.

2.3 GAMES IN EXTENSIVE FORM WITH COMPLETE INFORMATION

The Lady and the Tiger game is an example of a game in extensive form with complete information. A *game in extensive form with complete information* (GEFCI) has the following properties.

- It involves more than one individual.

- Each individual has more than one action.

- Once an action is chosen by each player, payoffs are determined for all players.

- Each individual is trying to maximize their payoff.

Note: The last property of the GEFCI is called the *assumption of rationality*. Given the choice between two or more possibilities, a player will make the decision that provides the highest payoff for them. This assumption is necessary for analysis in game theory, but it is hardly representative of real life.

Some definitions about GEFCI will also help keep things organized and explicit. A GEFCI has:

- A set $\mathbb{P}$ of players in the game.

- A set $\mathbb{N}$ of nodes, also known as the vertices of the game decision tree. The players from $\mathbb{P}$ are the labels of the nodes.

- A set $\mathbb{B}$ of actions. These actions label the edges of the game decision tree.

A *directed edge* is an arc connecting two vertices with a direction indicated by an arrow. A **root node** is any node that is not at the end of a move. A **terminal node** or a **leaf** is a node that is not the start of any move. A **path** is a sequence of moves such that at the end of a node of any move in the sequence is the start node of the next move in the sequence, excepting the last node in the sequence. A path is **complete** if it is not part of any longer path.

Exercise: Prove a complete path starts at the root and ends at a leaf.

A GEFCI has a function from the set of nonterminal nodes to the set of players. This is called the **labeling set**, and it defines which players choose a move at a given node. Each player has a payoff function from the set of complete paths to the real numbers. Players are labeled using the set $\{1, 2, ..., N\}$, and given payoffs π_i, where $i = 1, 2, .., n$, and $\pi_i \in \mathbb{R}$. A GEFCI must satisfy the following conditions.

1. A GEFCI must have exactly one root node.

2. If c is any node other than the root node, there is a unique path from the root node to c.

The consequences of condition 2 are that each node other than the root node is the end of exactly one move.

Exercise: Prove that each node other than the root node is the end of exactly one move.

In this treatment, we are only going to consider games of finite length called *finite horizon games*, which are GEFCI with finite complete paths.

2.4 BACKWARD INDUCTION

The assumption of rationality of players in the GEFCI implies the following.

- Suppose a player has a choice that includes two moves, a and b. If the payoffs from a and b are $\pi(a)$, $\pi(b)$ and $\pi(a) < \pi(b)$, then that player will choose b as their move.

- Since all players are following the principle of rationality, each player's choice(s) will affect every other choice's payoffs.

The implications of rationality help motivate a technique called "pruning the game decision tree," or "backward induction" [11].

Algorithm for Backward Induction:

1. Select a node n such that all moves from n end at a leaf. This can also be thought of as any node that is edge-distance 1 from a leaf.

2. Suppose n is labeled with player i, meaning it is player i's choice at that node. Find the move a such that a gives the greatest payoff to player i. Assume the payoff from a is unique.

3. Player i chooses a. Record this move as part of player i's strategy.

4. Delete all moves from the game decision tree that start at n. n now becomes a leaf, and assign to it the payoffs that a would have provided at the previous leaf.

5. Repeat steps 1–4 until there is only 1 node left in the game decision tree.

2.5   THE ULTIMATUM GAME

The Ultimatum Game (UG) consists of two players. Player 1 is given $L > 1$ dollars in 1 dollar coins, where L is finite. She must then decide on how to split the money between her and Player 2. She must offer at least 1 coin, but she could offer any amount between 1 and L inclusive to Player 2. If Player 2 accepts the offer, Player 1 and 2 split the coins according to the proposed offer and walk away with their money. If Player 2 rejects the offer, neither player gets to keep anything. We can build a game decision tree to outline all of the possible choices for Players 1 and 2, shown in Figure 2.4.

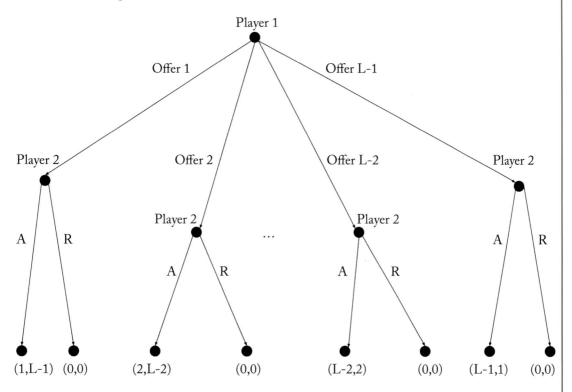

Figure 2.4: The game decision tree for the Ultimatum Game. A is accept the offer, R is reject the offer.

Notice that I restricted our attention to all offers between 1 and $L - 1$, since neither player is going to accept a situation in which they are left with no coins. If we use backward induction on this game, notice that for every choice available to Player 2, the payoff from accepting is better than the payoff from rejecting the offer. Based on this knowledge, when we cut off the bottom part of the tree and move the game to Player 1's decision, Player 1 will choose to offer the minimum value, 1, since that gives her the best payoff and she knows that Player 2 is rational and will accept any offer greater than zero.

2.5.1 THE ENHANCED ULTIMATUM GAME

The Enhanced Ultimatum Game (EUG), invented by Peter Taylor and the author, is a variation on the Ultimatum Game. It incorporates a *demand* into the game, and is played between two players as follows.

- Once again, there are L coins sitting on the table between the two players.

- Player 2 makes a demand of Player 1, stating "I will not take anything less than D!," where $0 < D < L$.

- Player 2 has a secret value, M, called the minimum accept. This is what Player 2 is actually willing to accept. We will restrict our attentions to $M \leq D$.

- There is a cost to giving a demand that is not equal the minimum accept value. The payoff to Player 2 is reduced by $c(D - M)$, $c > 0$.

- Player 1 must then respond to Player 2's demand (or not), and make a proposal, P, based on the information now available. Player 1 is trying to guess Player 2's M.

- If $P \geq M$, the deal goes through and the payoffs are $(L - P, P - c(D - M))$. If not, the payoffs are (0,0).

2.5.2 WHAT BACKWARD INDUCTION SAYS

The EUG has an element of probability involved, but we can construct the following scenario and extrapolate from there. Let $L = 3$, and let the cost, c, be zero. Consider Figure 2.5.

Player 2 begins by making a demand of 1 or 2. Player 1 then has to decide whether to offer 1 or 2. If Player 2 demands 1, Player 1 should offer 1 and the game is over with payoffs (2,1). If Player 2 demands 2, Player 1 can offer 1 or 2. If Player 1 offers 2, the payoffs are (1,2) and the game is over. However, if Player 1 offers 1, there is a probability $0 < p < 1$ (Player 1 can never be completely sure one way or the other) that Player 2 will accept the deal. If Player 1 believes that $p > 0.5$, he should offer 1 instead of 2.

However, that calculation is made moot by backward induction. If we prune the tree we end up with Player 2 making the final decision, where the possible payoffs are $(2,1)$ or $(2p, 1p)$. Since $1 > 1p$, the rational Player 2 will always demand 1. If we include a cost, $c > 0$, the result will still hold. The EUG has the same optimal strategy as the Ultimatum Game, assuming

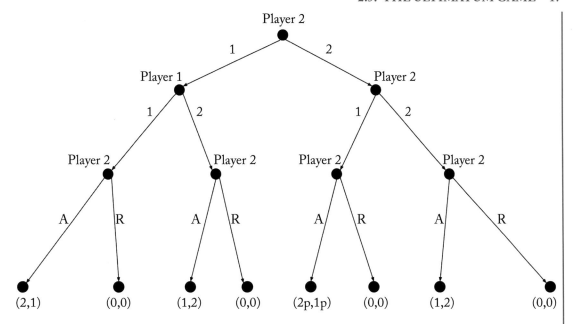

Figure 2.5: The game decision tree for the Enhanced Ultimatum Game, with $L = 3$. A is accept the offer, R is reject the offer.

players are rational, according to backward induction. It shouldn't be too hard to see this result can be extended for any integer value of $L > 3$.

2.5.3   BACKWARD INDUCTION IS WRONG ABOUT THIS ONE

In the Ultimatum Game, simulations done by the author show that under most conditions, agents that use the rational strategy (offer 1 if you are Player 1, accept anything if you are Player 2) are the most successful and take over a population that is not using the rational strategy. The optimal strategy of the EUG, however, based on a simulation run by the author, is not the same as the rational one.

Andrew's Model

In my model, $L = 20$ and I created 100 randomly generated agents that used arrays as their strategies. These arrays consisted of 22 values. The first 19 values were called the *response* values; they represented how an agent would play as Player 1 in the EUG. The response value at position 1 (position 0 was ignored) was what the agent would offer if a demand of 1 was received. The number at position 2 was the proposal if the agent received a demand of 2, and so on. The 20th and 21st values were, respectively, the minimum accept and the demand, and represented how the agent would play as Player 2. The 22nd value of the array represented how much money an agent currently had. For every agent, these values, except money, which started at 0, were randomly generated with the following restrictions.

- The proposal response could never be higher than the demand. By default, this meant that the value for every agent at position 1 in their array was 1.

- The minimum accept could never exceed the demand value.

- The minimum accept and demand were generated using a specialized distribution. A mean was set for each, and there was a 40% chance that the mean was used, a 40% chance that a value ±1 the mean was used, and a 20% chance that a value ±2 the mean was used. These choices were arbitrary and it remains to be seen if different distributions affect the results.

Figure 2.6 gives an example of what a typical array may look like, given that the mean minimum accept would be 8 and the mean minimum demand would be 12.

| 1 | 1 | 3 | 2 | 3 | 5 | 4 | 7 | 9 | 8 | 11 | 11 | 10 | 9 | 14 | 12 | 6 | 8 | 9 | 7 | 10 | 0 |

Figure 2.6: An example of an agent's array. The end value of zero denotes that this agent has made no money.

Once the agents were created, evolutionary computation [10] was used to find agents that performed well in the EUG. Without going into too much of the specifics, of the 100 agents, 2 were randomly selected to play a game. The part of the array that represented money was updated according to the result of the game. This was done 1 million times. Every 1,000 games, a 7 agent tournament selection was held. The top two performing agents were used as parents to produce two children with mutation, and those children replaced the worst performing agents in the tournament. 100 separate runs of this algorithm, with a mean minimum accept of 8 and a mean minimum demand of 10, with a cost of 0.5, produced the following results given in Figure 2.7, shown on the next page. The top graph is the average proposal response from the 100 top performing agents after 1 million games. Notice that there is a very strong linear trend that does not follow the backward induction solution to the game. The error bars are 99% confidence intervals. Also notice that in the demand vs. minimum accept graph only 21 out of 100 agents used a minimum accept of 1 (it doesn't show 21 dots due to overlap). The majority of the winning agents have their minimum accepts at different values, the largest group being between 4 and 8. I've only shown one set of parameters from the experiment, but I ran several using different initial conditions, varying the cost between 0 and 1, and interestingly, the results are strongly robust against all parameter changes. The linear trendline of the proposal as a response to the demand, $P = A \cdot D$, has A between 0.5 and 0.6 with an $R^2 > 0.99$ in *every case tested*. That means that on average, the most successful agents gave around 60% of the demand regardless of all other conditions. The average minimum accept of the 100 winners was 5.13 with a 95% confidence interval of ±0.791148361308, and the average demand was 9.93 with a 95% confidence interval

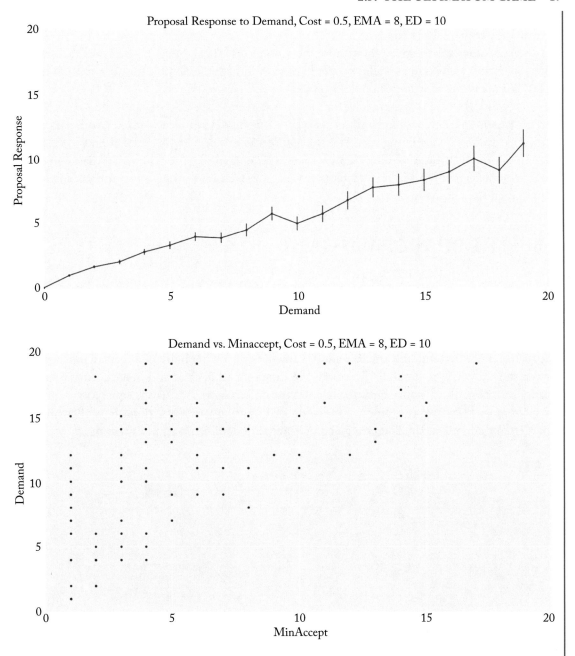

Figure 2.7: Some results of the EUG model study. The top graph is the average proposal response from the 100 top performing agents from the separate runs after 1 million games. The bottom graph is minimum accept vs. the demand of those top performing agents.

of ± 1.082993881771, which agrees with the proposal responses. Similar results were found in all other experimental settings.

What does this suggest about how the Enhanced Ultimatum Game should be played? First, some successful agents still used the backward induction solution for their minimum accepts, but *not one* used it to decide their proposals.

Second, the agents were able to figure out that even with a cost, most of the time it is better to demand more than your minimum accept. This model still needs work. More parameter testing, as well as determining an a distribution of minimum accepts and demands. Is a completely random distribution of minimum accepts and demands a better choice? My last comment on this model is that the successful agents are the ones who play more fairly (entirely subjective to me) than the ones who don't.

2.6   THE BOAT CRASH GAME

The following game was created by Spencer MacDonald, Will Heney, and Juliana Lebid. **Note:** The Boat Crash violates the condition that there is more than one individual involved in the game, but it represents an excellent opportunity to see the application of backward induction.

A lifeguard sitting in her watch-tower sees a small boat crash happen in the water. The lifeguard can see two children and two adults drowning. From her hardened years in the lifeguard tower, she knows approximately the amount of time that adults and children can survive in this sort of situation. The distance between her and the drowning individuals is equal to 9 time steps of swimming. However, the children and adults are floating laterally 2 time steps of swimming apart from each other. The diagram given in Figure 2.8 demonstrates the situation.

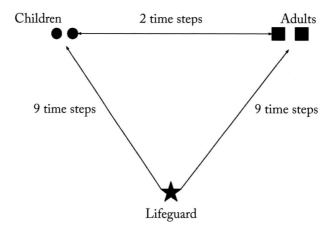

Figure 2.8: The Boat Crash Game, pictured.

The lifeguard knows that number of time steps the children can survive, defined as $s(c)$, is $s(c) = 12 - 1d6$, where $1d6$ is a representation of rolling a 6-sided die. So with uniform probability $s(c)$ could equal 11, 10, 9, 8, 7, and 6 time steps. The lifeguard also knows the number of time steps an adult could survive, defined as $s(a)$, is $s(a) = 31 - 1d6$, so the values that $s(a)$ could take on with uniform probability would be 30, 29, 28, 27, 26, and 25 time steps. Based on the picture, if the lifeguard swims the 9 time steps to save the children and finds that both of them are drowned, she can then swim 2 more time steps toward the adults' location to attempt to save their lives. The lifeguard can only save one person at a time, by swimming out there, retrieving a person, and dragging them back. What should the lifeguard do?

Our analysis requires some notation and assumptions.

- Let c be the value of saving one child's life to the lifeguard.

- Let a be the value of saving one adult's life to the lifeguard.

- People are useless at saving their own lives, and will remain where they are for the duration of the scenario unless the lifeguard comes to get them.

- The lifeguard is not slowed by swimming with another person.

- Once a lifeguard reaches a swimmer, the swimmer is no longer in danger of drowning.

- If the lifeguard arrives on the time step when the swimmer would drown, the swimmer does not drown.

- The lifeguard will not get tired, or risk drowning herself.

- Due to the fact that the game could take up to 29 time steps (9 out, possibly 2 lateral, 9 back, and 9 out again if anyone is still alive), the lifeguard will make 2 decisions.

Since there are probabilities involved, we need to determine the expected payoffs to the lifeguard from saving the children and the adults. Let's analyze what happens if she swims toward the children first.

- Each child has a probability of $\frac{1}{2}$ of being alive when the lifeguard arrives.

- The probability of both children being alive is $\frac{1}{4}$.

- The probability of 1 or 2 children being alive when the lifeguard arrives is $\frac{3}{4}$.

- Thus, the expected payoff to the lifeguard from swimming toward the children is $\frac{3}{4}c$.

- If both children are drowned, the lifeguard can swim 2 more time steps to rescue an adult, who will both still be alive at this point.

- Putting this together, if she swims toward the children first, $\frac{3}{4}$ of the time the lifeguard will rescue a child, and $\frac{1}{4}$ of the time she will rescue an adult. The expected payoff of this first round decision is $\pi = \frac{3}{4}c + \frac{1}{4}a$.

- In the second round, if the lifeguard had already saved a child, the probability that at least one adult survives is given by $\binom{2}{1}(\frac{1}{3})(\frac{2}{3}) = \frac{4}{9}$, and the probability both adults are still alive is $\frac{1}{9}$, so the probability of 1 or both adults still being alive is $\frac{5}{9}$. This is the probability that the lifeguard can rescue an adult, given that she already rescued a child in the first round.

- If the lifeguard does not rescue a child in the first round and swims to save an adult, she will not be able to rescue the second adult, so $\frac{3}{4}$ of the time the lifeguard will be able to rescue an adult with probability $\frac{5}{9}$. The expected payoff then becomes $(\frac{3}{4})(\frac{5}{9})a = \frac{5}{12}a$.

If the lifeguard chooses to go for the adults instead, she will rescue one for certain, and has a $\frac{1}{2}$ probability of rescuing the second adult, so the expected payoff would be $\pi = \frac{3}{2}a$.

We can use a decision tree to help us see all of the possibilities and payoffs, given in Figure 2.9.

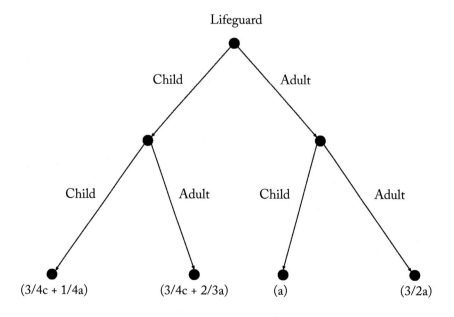

Figure 2.9: The decision tree of the Boat Crash Game.

If we let $c > \frac{10}{9}a$, then it becomes clear using backward induction that the lifeguard should choose to save a child first, and an adult second. If $c < \frac{10}{9}a$, then the lifeguard should attempt to save both adults, and leave the children to their fates; perhaps dolphins will come by and carry them to safety.

2.7   CONTINUOUS GAMES

The Lady and the Tiger game and the Ultimatum and Enhanced Ultimatum game are forms of games where players have a finite number of choices. There are many games where the choice a player can make is drawn from the domain of real numbers, and so represents an uncountable infinite spectrum from which to choose. This makes utilizing the technique of backward induction impossible: even if we have infinite time, the uncountability property of the real numbers still means we wouldn't be able to check every possibility. Enter calculus to save the day. The techniques in calculus allow us to use the continuity properties of the real numbers to find solutions without checking all of them. To keep things relatively simple, we restrict our attention to two player games, in which each player makes a single choice. The choice of each player and their opponent will affect their own payoffs. Those choices, assuming the players are still rational and want to pick the option that is best for them, will be called *best responses*. A best response is the choice that will maximize payoff of that player, assuming the other player makes the rational choice.

Imagine a game with two players, P1 and P2. Without loss of generality (which means we can interchange P1 and P2, and all of the attributes that go with each, and still get the same answers), P1 will make the first choice, $a \in A$, where $A \subset \mathbb{R}$. Then P2 can see P1's choice, and will also make a choice $b \in B$, where $B \subset \mathbb{R}$. After the two decisions have been made, the payoffs to P1 and P2 are calculated, $\pi_{P1}(a, b)$ and $\pi_{P2}(a, b)$. Technically, this is still a game in extensive form with complete information, and it shouldn't be too hard to satisfy yourself that it fulfills all the conditions given in the section describing GEFCI. However, even though backward induction does not work in this regard, a similar concept will help us arrive at a solution.

Start by examining P2's decision, since that is the last move made in the game. With the rationality assumption, P2 will observe P1's decision, a, and then maximize the function $\pi_{P2}(a, b)$, with a fixed. Assuming the payoff function is continuous on the interval we are concerned with, partial differentiation with respect to b of $\pi_{P2}(a, b)$ will determine the value of b that maximizes P2's payoff, and that value $b = m(a)$ will be some function of that fixed a.

The function $b = m(a)$ is called the *best response function*, and it gives the best choice for P2 based on the decision P1 will make. P1 will take that information into account, just like P1 would in regular backward induction, and make a choice a. P1's payoff will then be $\pi_{P1}(a, m(a))$, which means that P1's payoff function has only one variable. Assuming that payoff function is continuous on the interval we are concerned with, we can use calculus again to find the value of a that will maximize the payoff to P1.

2.7.1   TWO MODELS OF THE STACKELBERG DUOPOLOGY

The First Model

In a *duopoly*, a certain good is produced and sold by two companies, ACME 1 and ACME 2. Each company wants to choose a level of production that will maximize its profit. Without loss of generality, let us assume that ACME 1 chooses its own level of production first, and ACME

2 chooses its level of production after observing the choice made by ACME 1. We need to make some assumptions to start.

- ACME 1 chooses first, and ACME 2 chooses second, after observing ACME 1.

- ACME 1 produces a quantity a of the product, and ACME 2 produces a quantity b of the product.

- The total quantity on the market produced by these two companies is $q = a + b$.

- The price of the product is based on the quantity available in the market, $p = f(q)$, and everything that is produced is sold.

Now let us suppose that ACME 1's cost to produce a of the good is $c_1(a)$, and the ACME 2's cost to produce b of the good is $c_2(b)$. The profit of the companies, $\pi_{ACME1}(a, b)$ and $\pi_{ACME2}(a, b)$, are given below:

$$\pi_{ACME1}(a, b) = f(a + b)a - c_1(a) \text{ and } \pi_{ACME1}(a, b) = f(a + b)b - c_2(b).$$

This first model also requires the following assumptions.

1. Price falls linearly with the total production such that $p = \alpha - \beta(a + b), \alpha, \beta > 0$. Notice that this allows the price to be negative.

2. ACME 1 and ACME 2 have the same cost to produce the good, so $c_1(a) = ca$ and $c_2(b) = cb, c > 0$.

3. $\alpha > c$. This ensures that the cost of producing the good is lower than the sale price. Otherwise there would be no point in producing the good at all (in this scenario).

4. $a, b \in \mathbb{R}$.

With these assumptions we can now investigate the following question: Is it better to choose your production levels first, or to observe your competitor's production level and then make a decision?

The new assumptions allow us to construct the following payoff functions for ACME 1 and ACME 2:

$$\pi_{ACME1}(a, b) = f(a + b)a - ca = (\alpha - \beta(a + b) - c)a = (\alpha - \beta b - c)a - \beta a^2$$

$$\pi_{ACME2}(a, b) = f(a + b)b - cb = (\alpha - \beta(a + b) - c)b = (\alpha - \beta a - c)b - \beta b^2.$$

Since ACME 1 chooses first, we can find ACME 2's best response $b = m(a)$. First we fix a, meaning that $\pi_{ACME2}(a, b)$ is a parabola opening downward, and we can find the maximum using:

$$\frac{\partial \pi_{ACME2}}{\partial b}(a, b) = \alpha - \beta a - c - 2\beta b.$$

If we set that quantity equal to zero and solve for b we find that

$$b = \frac{\alpha - \beta a - c}{2\beta} = m(a).$$

Using $b = m(a)$, we can then find ACME 1's best response to ACME 2's choice:

$$\pi_{ACME1}(a, m(a)) = \frac{\alpha - c}{2}a - \frac{\beta}{2}a^2.$$

This payoff function is also a parabola opening downward, and its maximum can be found with:

$$\frac{\partial \pi_{ACME1}}{\partial a}(a, m(a)) = \frac{\alpha - c}{2} - \beta a.$$

Setting this equal to zero and solving for a gives $a = \frac{\alpha - c}{2\beta}$. Thus, ACME 1 obtains its maximum profit at $a^* = \frac{\alpha - c}{2\beta}$. Since ACME 2 can observe a^*, its own choice must be

$$b^* = m(a^*) = \frac{\alpha - \beta(\frac{\alpha - c}{2\beta}) - c}{2\beta} = \frac{\alpha - c}{4\beta}.$$

The profits from the decisions a^* and b^* will be:

$$\pi_{ACME1}(a^*, b^*) = \frac{(\alpha - c)^2}{8\beta}$$

$$\pi_{ACME2}(a^*, b^*) = \frac{(\alpha - c)^2}{16\beta}.$$

We can see that ACME 1 has twice the profit of ACME 2. In this model, it is better to be the company that chooses the production level first.

The Second Model

In the second version of the duopoly, we now restrict our attention to production levels and prices that will make at least a little more sense. Assume the production levels $a, b \geq 0$, which forces $p \geq 0$, and now let the price function be defined piecewise in the following manner:

$$p(a, b) = \begin{cases} \alpha - \beta(a + b) & \text{if } a + b < \frac{\alpha}{\beta} \\ 0 & \text{if } a + b \geq \frac{\alpha}{\beta}. \end{cases}$$

We let all of the assumptions in the first model remain the same in the second model, excepting the assumptions already stated earlier. The question once again becomes: Is it better to choose your production levels first, or to observe your competitor's production level and then make a decision? The profit function for each company is now a piecewise function, defined below:

$$\pi_{ACME1}(a,b) = f(a+b)a - ca = \begin{cases} (\alpha - \beta(a+b) - c)a & \text{if } 0 \leq a+b < \frac{\alpha}{\beta} \\ -ca & \text{if } a+b \geq \frac{\alpha}{\beta} \end{cases}$$

$$\pi_{ACME2}(a,b) = f(a+b)b - cb = \begin{cases} (\alpha - \beta(a+b) - c)b & \text{if } 0 \leq a+b < \frac{\alpha}{\beta} \\ -cb & \text{if } a+b \geq \frac{\alpha}{\beta}. \end{cases}$$

Our approach remains the same as in the first model to determine the best response of each company. First, we find $b = m(a)$. If ACME 1 produces $a \geq \frac{\alpha}{\beta}$, ACME 2 should not produce anything. If ACME 1 wants to drive ACME 2 right out of the market, the level of production necessary to drive the price equal to the cost, $c = \alpha - \beta a$ is $a = \frac{\alpha - c}{\beta}$. If $a \geq \frac{\alpha - c}{\beta}$, ACME 2's best response is to produce nothing.

If, for whatever reason, $s < \frac{\alpha - c}{\beta}$, this leaves an opening for ACME 2. ACME 2's payoff function can now be defined as

$$\pi_{ACME2}(a,b) = f(a+b)b - cb = \begin{cases} (\alpha - \beta(a+b) - c)b & \text{if } 0 \leq a+b < \frac{\alpha - \beta a}{\beta} \\ -cb & \text{if } a+b \geq \frac{\alpha - \beta a}{\beta}. \end{cases}$$

Notice the change in the conditions of the piecewise function. These are a direct result of the price function given in the beginning of the description of this version of the model and the fact that $s < \frac{\alpha - c}{\beta}$. I leave it to you to verify the inequality is correct.

To find the maximum of $\pi_{ACME2}(a,b)$ we once again take the partial derivative, set it equal to zero, and solve for b. This gives

$$b = \frac{\alpha - \beta a - c}{2\beta}.$$

We can then define ACME 2's best response as

$$b = m(a) = \begin{cases} \frac{\alpha - \beta a - c}{2\beta} & \text{if } 0 < a < \frac{\alpha - c}{\beta} \\ 0 & \text{if } a \geq \frac{\alpha - c}{\beta}. \end{cases}$$

Using this, we can find the payoff to ACME 1, $\pi_{ACME1}(a, m(a))$. Note that for $0 < a < \frac{\alpha - c}{\beta}$

$$a + m(a) = a + \frac{\alpha - \beta a - c}{2\beta} < \alpha + \beta(\frac{\alpha - c}{\beta}) < \frac{\alpha}{\beta}.$$

Therefore, for $0 < a < \frac{\alpha - c}{\beta}$

$$\pi_{ACME1}(a, m(a)) = (\alpha - \beta(a + \frac{\alpha - \beta a - c}{2\beta}) - c)a = (\frac{\alpha - c}{2})a - \frac{\beta}{2}a^2.$$

We do not need to consider the case when $a \geq \frac{\alpha - c}{\beta}$, since the cost is greater than or equal to than the price in that region.

Using the same techniques as before, if we consider the interval $\left[0, \frac{\alpha - c}{\beta}\right)$ and the payoff function $\pi_{ACME1}(a, m(a))$, if we use differentiation we will find that the maximum of the payoff function occurs at $a^* = \frac{\alpha - c}{2\beta}$, and that implies that $b^* = \frac{\alpha - c}{4\beta}$. Once again, the profit of ACME 1 doubles that of ACME 2 if ACME 1 chooses first, so it is better to be the company that sets production levels. Assuming both companies are rational.

2.8   THE FAILINGS OF BACKWARD INDUCTION

Backward induction fails when the payoffs given by distinct moves from a single node are equal. If this should happen at any point in the game, all of the players will not know which decision the deciding player will make and this makes further analysis of the problem impossible with this technique. The requirement that moves have unique payoffs limits backward induction, although it is not unreasonable to imagine that there aren't many cases where the payoff is *exactly* the same for two distinct choices.

The main concern with backward induction as a technique to solving games where the players take turns making moves is the assumption of *Common Knowledge of Rationality*. This is the idea that the players are rational, and they know the other players are rational, and they know that the other players know that they are rational. Carrying around all of those beliefs about what the other players are going to do, especially if a game requires several moves from each player, requires that each player hold several layers of beliefs, and this makes backward induction difficult to apply in complex situations that involve real people.

2.9   EXERCISES

1. Refer back to the Lady and the Tiger Game. What is the conclusion of the game using backward induction if $m < -D$?

2. Consider the Boat Crash Game.

    (a) If the value of the life of the adults and the children are equal to the lifeguard, what strategy should the lifeguard employ?

    (b) If the lifeguard can swim $41 - 1d6$ rounds, what would the value of her own life have to be such that she decides to make a second trip to the survivors of the boat crash?

    (c) If we change the survivorship function of the children to be $s(c) = 11 - 1D6$, how does this affect the lifeguard's strategy?

    (d) If we change the survivorship function of the adults to be $s(a) = 34 - 1D6$, how does this affect the lifeguard's strategy?

3. Numerical simulations are fine for getting an idea of how a game should be played, but the elements of stochasticity mean we are never quite sure of the result. Consider a simplified version of the EUG. In this version, we have a finite population, and two types of people exist. Those with high minimum accepts, H, and those with low minimum accepts, L. Recall that while a person has a fixed minimum accept, they may demand any value.

(a) Assuming no one knows anything about the distribution of H's and L's, but the values are known, is there a best strategy an H person can employ? What about an L person? What should the proposer do when he or she receives a demand?

(b) Now assume that the game has been played for a while with this population, and the distribution of the demands of H's and L's are known. For our purpose, let's assume both of those distributions are half-Gaussian (see the associated example picture). Characterize all distributions such that an L-person could signal that they are actually an H-person, called *dishonest signalling*, and the payoff to the L-person would still be positive. In this family of scenarios, what should an H person do? What should the proposer do when he or she receives a demand?

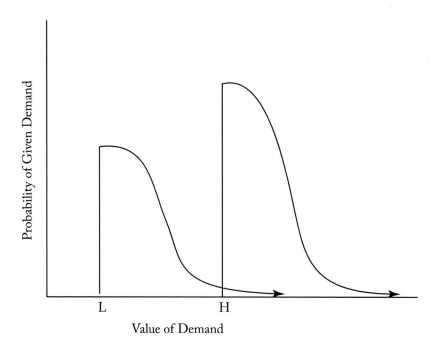

Figure 2.10: Demand: probability vs. value.

(c) Now characterize the distributions where the payoff of dishonest signalling is negative. What should an H-person do in this case? What should the proposer do when he or she receives a demand?

4. Another simplified version of the EUG is as follows. A population of Acceptors is given a minimum accept, x, not by choice, but by Nature according to the Gaussian distribution, with a mean of $\bar{x}$ and some variance σ^2. It is well known that these Acceptors make their demands, y, using the function $y = ax + b$. If you were a Proposer, playing against this population, you would receive the demand y, and would have to come up with a proposal, z.

   (a) Assuming the demand, y, is sent across a noiseless channel, what is the calculation you need to make to find the original minimum accept, x?

   (b) Now assume the demand is sent across a noisy channel, such that the Proposer receives $\hat{y} = y + \epsilon$, where $\epsilon = N(0, \sigma^2)$. Is there a way to figure out x? If not, what value z should be used for your offer?

5. Come up with your own situation that can be described as a game that will use backward induction to solve. Keep things simple, and be sure to describe the players, their possible actions, the order of the players decision's, and more importantly, find the solution to the game. Creativity is appreciated.

CHAPTER 3

Games in Normal Form and the Nash Equilibrium

3.1 INTRODUCTION AND DEFINITIONS

Games in extensive form with complete information can be used to think about many situations and help us analyze how to approach a given decision. However, GEFCI are limited because they can never represent a situation in which both players make their decisions at the same time. Unlike the games in extensive form with complete information, in which players take turns making moves, a *game in normal form* (GNF) consists of players making decisions *simultaneously*. A GNF:

- Has a finite set of players $\mathbb{P} = \{1, ..., n\}$.

- For each $i \in \mathbb{P}$, S_i is the set of available strategies to each player.

  Note: We can define the *strategy* of a player p, S_p, as the move p will make for every situation they encounter in the game. The size of S_p, $|S_p|$ is given by the available choices to p in the game. Using this definition, we can also define $\mathbb{S} = S_1 \times S_2 \times \ldots \times S_n$ as the complete space of strategies for a given n player game.

- An element of $\mathbb{S}$ is an n-tuple $(s_1, ..., s_n)$, where each s_i is a strategy chosen from S_i.

- The n-tuple $(s_1, ..., s_n)$ is called a *strategy profile*, and represents the choice of strategy of each the players participating in the game.

- For each $i \in \mathbb{P}$, there exists a payoff function $\pi_i : \mathbb{S} \to \mathbb{R}$.

Consider the game with two players. If player 1 has m strategies and player 2 has n strategies, then the game in normal form can be represented as an $m \times n$ matrix of ordered pairs of payoffs. This chapter has a few examples to illustrate the concepts being described, try to describe the set of players, their available strategies, and a couple of strategy profiles for each example. This will help you keep the technical parts of this chapter understandable.

3.2 THE STAG HUNT

The story of the Stag Hunt, initially told by Jean-Jacques Rousseau, in *A Discourse on Inequality* in the 18th century, is as follows.

"If it was a matter of hunting a deer, everyone well realized that he must remain faithful to his post; but if a hare happened to pass within reach of one of them, we cannot doubt that he would have gone off in pursuit of it without scruple..."

Another way to think of this is two hunters go out and agree to occupy opposite sides of a game track. The hunters can't see each other, but when large prey comes by, such as a stag, one hunter can attempt to kill it. If the stag were to run away, it would run into the other hunter waiting for it, thus increasing the chances that the two hunters could bring the stag down. Hunting a stag is good, likely to provide for the hunters and their families for a while. However, if one of the hunters abandons their post to hunt down easier prey, say a hare, that hunter probably will get the hare and eat for a day, but they leave the other hunter without help, making it unlikely that the stag will be brought down. The other hunter will go hungry.

Unlike the the Prisoner's Dilemma, the Stag Hunt does not have a sucker or temptation payoff. The payoff of defecting (hunting hare) if your opponent cooperates (hunting stag) is not better than if you both hunted stag. David Hume and Thomas Hobbes also have discussed their own versions of this concept, but the man in recent history to have really nailed down the Stag Hunt is Brian Skyrms. His book *The Stag Hunt and Evolution of Social Structure* is an excellent read, and he has since written several papers on the topic since the release of the book. The Prisoner's Dilemma has been used as a model for social interaction for many years, but Skyrms, among others, has given strong arguments that the Stag Hunt is better at representing many of those social situations. We can construct the Stag Hunt as a game in normal form with the following conditions.

- There are two players.

- Each player has the available strategy *hunt stag* or *hunt hare*.

- The payoffs for both players are given in the following matrix:

		Player 2	
		HS	HH
Player 1	HS	(5,5)	(0,2)
	HH	(2,0)	(2,2)

Similar to games in extensive form, payoffs are listed as ordered pairs (or n-tuples if there are more two players), with the first term being the payoff to Player 1, the second to Player 2, and so on. So in our version of the Stag Hunt, if both players hunt stag, they receive a payoff of 5. We can generalize the Stag Hunt by creating the payoff matrix:

		Player 2	
		HS	HH
Player 1	HS	(a,a)	(c,b)
	HH	(b,c)	(d,d)

If $a \geq b \geq d \geq c$, then this is called a Generalized Stag Hunt. If the conditions that $2a \geq b + c$ and $b \geq a \geq d \geq c$ are satisfied, this payoff matrix represents a Prisoner's Dilemma.

3.3 DOMINATED STRATEGIES

For a game in normal form, let s_i and s_j be two of player p's strategies ($s_i, s_j \in S_p$).

- We say s_i *strictly dominates* s_j if, for every choice of strategies by the other n-1 players of the game, $\pi_p(s_i) > \pi_p(s_j)$.

- We say s_i *weakly dominates* s_j if, for every choice of strategies by the other n-1 players of the game, $\pi_p(s_i) \geq \pi_p(s_j)$.

The game in normal form still assumes rationality of all players and this implies that if player p's strategy s_i weakly (or strictly) dominates s_j, then p will not choose s_j.

3.3.1 ITERATED ELIMINATION OF DOMINATED STRATEGIES

Recall that backward induction helped us solve games in extensive form with complete information by eliminating moves that yielded lower payoffs. The removal of strategies that yielded lower payoffs reduced the size of the game, assuming that different moves yielded unique payoffs from a given node. Games in normal form follow a similar process. The assumption of rationality of players implies that dominated strategies will not be used by a player. Thus, we can remove a dominated strategy from a game in normal form, and we are often left with a smaller game to analyze.

If the game reduces down to one strategy for a given player, we call that the *dominant strategy*. If there is a single dominant strategy for every player in the game, the strategy profile $(s_1^*, \ldots, s_n^*)$ is called a *dominant strategy equilibrium*.

Fact: Iterated elimination of *strictly* dominated strategies (IEDS) produces the same reduced game, regardless of the order in which dominated strategies are removed from the game. However, order of elimination can matter when removing *weakly* dominated strategies.

3.4 THE NACHO GAME

The Nacho Game was created by Brandon Chai and Vincent Dragnea. As they put it, this game was born at the pub, over a plate of nachos. When you share a plate of nachos with a friend, you have the option of eating nachos at an enjoyable, slower pace, or eating the nachos at a quicker, less enjoyable pace that ensures you eat more of them than your friend. This can be modeled as a two-player game in normal form.

- For the ease of analysis, we limit the speed at which a player can eat nachos: slow and fast.

- At slow speed, a player eats 5 nachos per minute, and gets a payoff of 1 for each nacho eaten.

- At fast speed, a player eats 10 nachos per minute, and gets a payoff of $\frac{4}{5}$ for each nacho eaten.

- Both players make one decision, the speed at which to eat the nachos.

- A plate of nachos has n nachos to be consumed at the start of the game.

The payoffs for the Nacho Game are as follows:

		Player 2	
		Eat Slow	Eat Fast
Player 1	Eat Slow	$(\frac{n}{2}, \frac{n}{2})$	$(\frac{n}{3}, \frac{8n}{15})$
	Eat Fast	$(\frac{8n}{15}, \frac{n}{3})$	$(\frac{2n}{5}, \frac{2n}{5})$

If we examine the payoffs, this game satisfies the conditions of a Prisoner's Dilemma! In terms of dominated strategies, we can see that choosing to eat nachos at a fast pace (defect) always provides a larger payoff than choosing to eat nachos at a slow pace (cooperate), regardless of what decision the other player makes. Thus, the solution to the Prisoner's Dilemma, based on IEDS, is that both players should defect.

3.4.1   THE NACHO GAME WITH K PLAYERS

Unlike the Prisoner's Dilemma (not including its many, many variants), the Nacho Game allows for more than two players to participate. Indeed, picture a round table full of hungry undergraduate students staring at a plate of fresh, hot, cheese-covered nachos, along with all your favorite toppings. We will start by analyzing the game with three players, and generalize from there. A modified payoff matrix is necessary here.

	(Fast,Fast)	(Fast,Slow)	(Slow,Fast)	(Slow,Slow)
Fast	$(\frac{4n}{15}, \frac{4n}{15}, \frac{4n}{15})$	$(\frac{8n}{25}, \frac{8n}{25}, \frac{n}{5})$	$(\frac{8n}{25}, \frac{n}{5}, \frac{8n}{25})$	$(\frac{2n}{5}, \frac{n}{4}, \frac{n}{4})$
Slow	$(\frac{n}{5}, \frac{8n}{25}, \frac{8n}{25})$	$(\frac{n}{4}, \frac{2n}{5}, \frac{n}{4})$	$(\frac{n}{4}, \frac{n}{4}, \frac{2n}{5})$	$(\frac{n}{3}, \frac{n}{3}, \frac{n}{3})$

This modified payoff matrix shows the payoffs to all three players. The rows are the choices made by Player 1 (without loss of generality, this could be Player 2 or 3). The columns represent the choices made by the other players, and the ordered triplets are the respective payoffs. If we use IEDS, once again we can see that it is in the best interest of Player 1 to eat fast, regardless of what choices the other players make.

We can generalize, and reduce, this payoff matrix for a single player regardless of how many players participate in the Nacho Game. Let k be the number of players, and let h be the number of players - not including yourself - that decide to eat fast. Then the payoffs to Player 1 are

	Payoff
Fast	$\frac{2n}{k+h+1} \cdot \frac{4}{5}$
Slow	$\frac{n}{k+h}$

Is it ever more profitable to eat slowly? We can check the conditions under which the slow payoff is greater than the fast payoff.

$$\frac{n}{k+h} > \frac{2n}{k+h+1} \cdot \frac{4}{5} \tag{3.1}$$

$$\frac{5}{k+h} > \frac{8}{k+h+1} \tag{3.2}$$

$$5k + 5h + 5 > 8k + 8h \tag{3.3}$$

$$5 > 3(k+h). \tag{3.4}$$

If $k = 1$, the equation is true. So it is better to eat slow when you are eating a plate of nachos by yourself (no comment on the sadness of this endeavour). However, if $k > 1 \rightarrow h >= 0$, $3(k+h)$ is never less than 5, so under no conditions is the slow payoff greater than the fast payoff with more than 1 person playing. Thus, we have shown that the eat fast strategy strictly dominates the eat slow strategy.

3.5   NASH EQUILIBRIA

Consider a game in normal form with n players, with strategy sets $S_1, S_2, \ldots, S_n$ and payoffs $\pi_1, \pi_2, \ldots, \pi_n$. A *Nash equilibrium* is a strategy profile $(s_1^*, s_2^*, \ldots, s_n^*)$ with the following property: if any player changes his strategy, his own payoff will not increase. The strategy profile $(s_1^*, s_2^*, \ldots, s_n^*)$ is a Nash equilibrium if:

- For all $s_1 \in S_1, \pi_1(s_1^*, s_2^*, \ldots, s_n^*) \geq \pi_1(s_1, s_2^*, \ldots, s_n^*)$

- For all $s_2 \in S_2, \pi_2(s_1^*, s_2^*, \ldots, s_n^*) \geq \pi_2(s_1^*, s_2, \ldots, s_n^*)$

$$\vdots$$

- For all $s_n \in S_n, \pi_n(s_1^*, s_2^*, \ldots, s_n^*) \geq \pi_n(s_1^*, s_2^*, \ldots, s_n)$.

    The strategy profile $(s_1^*, s_2^*, \ldots, s_n^*)$ is a *strict* Nash equilibrium if:

- For all $s_1 \neq s_1^*, \pi_1(s_1^*, s_2^*, \ldots, s_n^*) > \pi_1(s_1, s_2^*, \ldots, s_n^*)$

- For all $s_2 \neq s_2^*, \pi_2(s_1^*, s_2^*, \ldots, s_n^*) > \pi_2(s_1^*, s_2, \ldots, s_n^*)$

$$\vdots$$

- For all $s_n \neq s_n^*, \pi_n(s_1^*, s_2^*, \ldots, s_n^*) > \pi_n(s_1^*, s_2^*, \ldots, s_n)$

    John Nash Jr. is responsible for introducing the Nash equilibrium in 1950 and 1951, and the concept has since appeared in almost every field that has applied game theory. The following theorem guarantees that, for a certain class of games, there exists a Nash equilibrium.

Theorem 3.1   *Every game with a finite number of players, each of whom has a finite set of strategies, has at least one Nash equilibrium.*

The proof of this is well beyond the scope of this course, but for the interested reader, you can find an excellent tutorial on the proof here: http://www.cs.ubc.ca/~jiang/papers/NashReport.pdf.

The Nash equilibrium (NE) is one of the most important concepts in classical game theory. You will notice that we have been finding the NE in every game I have introduced, although I have not specified it as such. If you go back and look over the Nacho Game, the solution to that game is the NE, as it has been defined above. The NE of the Prisoner's Dilemma is Defect, and the Stag Hunt has two NE: (HS,HS) and (HH,HH). From now on when analyzing games we will be searching for the NE when appropriate.

3.5.1   FINDING THE NE BY IEDS

The following two theorems demonstrate the utility of IEDS for finding NE.

Theorem 3.2   *Suppose we do iterated elimnation of weakly dominated strategies on a game G, where G is a GNF. Let H be the reduced game. Then:*

1. *each NE of H is also a NE of G and*

2. *if H has one strategy, s_i^* for each player, then the strategy profile $(s_1^*, s_2^*, \ldots, s_n^*)$ is a NE of G.*

Theorem 3.3   *Suppose we do IEDS, where the strategies are strictly dominated in a game G, G is a GNF. Then:*

1. *strategies can be eliminated in any order, it will result in the same reduced game H;*

2. *each eliminated strategy is not part of the NE of G;*

3. *each NE of H is a NE of G; and*

4. *if H only has s_i^* for each player, then the strategy profile $(s_1^*, s_2^*, \ldots, s_n^*)$ is a strict NE of G, and there are no other NE.*

The proof of both of these theorems can be found in [11], although it is not too difficult to see why both of these theorems are true.

3.5.2   IEDS PROCESS

The IEDS process for two players is actually fairly simple, and can easily be generalized to a game with n players. Considering the two player game, the easiest way to remember how to eliminate dominated strategies is with the following two ideas.

- If Player 1 has a dominated strategy by some other strategy, that will be shown by comparing the first element of each pair in two **rows**. Each element of the dominant strategy will be greater than each element of the dominated strategy, and you can elminate that entire **row** from the matrix.

- If Player 2 has a dominated strategy by some other strategy, that will be shown by comparing the second element of each pair in two **columns**. Each element of the dominant strategy will be greater than each element of the dominated strategy, and you can eliminate that entire **column** from the matrix.

- Bottom line: you eliminate rows by comparing first terms in the pairs, and you eliminate columns by comparing second terms. The order of elimination is irrelevant when finding Nash equilibria by iterated elimination of strictly dominated strategies, but it can matter when some or all of the eliminated strategies are weakly dominated.

For example,

$$G = \begin{pmatrix} (15,2) & (0,3) & (8,2) \\ (4,1) & (4,4) & (7,3) \\ (-2,8) & (3,9) & (7,-2) \end{pmatrix}.$$

Take a look at the middle and right columns of G. Notice that $3 > 2, 4 > 3, 9 > -2$, so the middle column dominates the right column, and we can delete the right column, getting

$$H = \begin{pmatrix} (15,2) & (0,3) \\ (4,1) & (4,4) \\ (-2,8) & (3,9) \end{pmatrix}.$$

Now the middle and bottom row of H can be compared, and since $4 > -2, 4 > 3$, we can see that the middle row dominates the bottom, so we delete the bottom row.

$$H = \begin{pmatrix} (15,2) & (0,3) \\ (4,1) & (4,4) \end{pmatrix}.$$

If we compare the left and right column, $3 > 2, 4 > 1$, thus we delete the left column.

$$H = \begin{pmatrix} (0,3) \\ (4,4) \end{pmatrix}.$$

And finally, the bottom row dominates the top row, so the Nash equilibrium of this game is the strategies which give the payoffs (4,4).

3.6   THE VACCINATION GAME

The Vaccination Game, first defined by [12], is an excellent combination of game theory and mathematical biology. If a group of people face the possibility of catching an infectious disease and have the option getting vaccinated with a completely effective vaccine, how many will choose to be vaccinated? The authors use a static model where individuals have to choose whether or not to be vaccinated. The group of people in the model consider the costs and benefits of those decisions by making assumptions about who else will get vaccinated. Those costs can be thought of as monetary, psychological, and health related (perhaps as an adverse reaction to the vaccine). The following definitions will be necessary to discuss this model.

- Let c_i be the cost to Player i to get vaccinated.

- Let L_i be the loss to i if they catch the disease. We assume that an individual can still catch the disease from the background even if they do not encounter another person for the duration of the disease.

- Define p_i to be the probability of catching the disease even if no one else has it.

- Define r to be the probability that an infected person will infect someone who is not vaccinated, thus rp_i is the chance of catching the disease from a non-human source and then infecting another susceptible person. Denote rp_i as q_i.

- Assume a person who is vaccinated cannot transfer the disease.

- Y_i is i's initial income or welfare.

Let us consider the two person case, where V is the choice to vaccinate, and NV is the choice to not vaccinate.

Table 3.1: Vaccination.

	V	NV
V	$Y_1 - c_1, Y_2 - c_2$	$Y_1 - c_1, Y_2 - p_2 L_2$
NV	$Y_1 - p_1 L_1, Y_2 - c_2$	$Y_1 - p_1 L_1 - (1 - p_1)q_2 L_1, Y_2 - p_2 L_2 - (1 - p_2)q_1 L_2$

From this payoff matrix in Table 3.1, we can do the following analysis.

1. When $c_i < p_i L_i$, for $i = 1, 2$, then (V,V) is a Nash equilibrium, as these strategies dominate all others.

2. For $p_1 L_1 < c_1$ and $c_2 < p_2 L_2 + (1 - p_2)q_1 L_2$, (NV,V) is a NE, and (V,NV) is an equilibrium if 1 and 2 are exchanged in the inequalities.

3. For $p_i < c_i < p_i L_i + (1 - p_i)q_j L_i$ for $i, j = 1, 2$, then both (NV,V) and (V,NV) are NE.

4. For $p_i L_i + (1 - p_i)q_j L_i < c_i$, for $i, j = 1, 2$, then (NV, NV) is the NE.

Note: Another paper, *The Theory of Vaccination and Games*, by [26], also uses game theory to analyze vaccination strategies.

3.6.1   THE N-PLAYER VACCINATION GAME

Consider the game with N people rather than just two. Define $R_i(K)$ to be the risk to individual i of infection when she is not vaccinated and those agents in the set $\{K\}$ are vaccinated. With this notation, let's restate the results of the two person vaccination game:

1. When $c_i < R_i(j)L_i$, for $i = 1, 2$, then (V,V) is a Nash equilibrium.

2. For $R_1(2)L_1 < c_1$ and $c_2 < R_2(\emptyset)L_2$, (NV,V) is a NE, and (V,NV) is an equilibrium if 1 and 2 are exchanged in the inequalities.

3. For $R_i(j)L_i < c_i < R_i(\emptyset)L_i$ for $i, j = 1, 2$, then both (NV,V) and (V,NV) are NE.

4. For $R_i(\emptyset)L_i < c_i$, for $i, j = 1, 2$, then (NV, NV) is the NE.

This suggests that there is a general pattern, given in the proposition below.

Proposition 3.4   *Let there be N people exposed to an infectious disease, with a probability r of catching this from an infected person and a probability p of catching it from a non-human host. c is the cost of vaccination and L is the loss from catching the disease. $R(\emptyset) = R(0)$ is the probability of a non-vaccinated person catching the disease if no one is vaccinated, and $R(k)$ is the probability of a non-vaccinated person catching the disease if k are vaccinated. Then at the Nash equilibria the number of people vaccinated is as follows: for $R(j)L < c < R(j-1)L$ there are j people vaccinated, $N - j$ not vaccinated. For $c < R(N-1)L$, everyone is vaccinated, and for $R(0)L < c$, no one is vaccinated.*

The proof of this proposition can be found in the paper by Heal and Kunreather.

3.7   EXERCISES

1. Consider the Vaccination Game, and let's impose a structure on the population of N people. A *small world network* is one in which the population is divided into small clusters of highly connected people, like a family, or a classroom of young students, or people who work in the same office every day. These clusters of people are weakly connected to other clusters. So a family might be connected to another family down the street by virtue of the fact of the friendship between a single child from each family. Or two offices are connected by the custodian that cleans each office at the end of the day. There are many varieties of this sort of small world network. Picture a small world network on N people, where there are n networks, and on average each network contains $\frac{N}{n}$ people, give or take a few. In each network, everyone is completely connected to everyone else, or near enough that it makes no difference. Each network is connected to at least one other network by one person. Modify Proposition 3.4 such that it takes the small world network into account, and prove that it is the optimal strategy.

2. **The tragedy of the commons.** The Nacho Game is an example of a class of situations called the tragedy of the commons.[1] Here is another version. There are s students who have to do an assignment question. The students know that the professor is likely to not notice that the students have copied at least a few of their answers from each other, or the Internet (and shame on those students). In fact, the students know that the professor is likely to let a total as of these questions get by. The ith student has a choice of two strategies.

   (a) The responsible strategy: cheat on a answers.

   (b) The irresponsible strategy: cheat on $a + 1$ answers.

   Each answer that is gained by cheating gives a payoff of $p > 0$ to the student. However, each student who cheats on $a + 1$ answers imposes a cost of getting caught $c > 0$ on the community of students, because of the professor will start to notice and dish out penalties accordingly. The cost is shared equally by the s students, because the professor can't know who cheated more or less than others. Assume $\frac{c}{s} < p < c$. Thus, the cost of cheating on one more answer is greater than the profit from the cheating, but each student's share of the cost is less than the profit.

   (a) Show that for each student, cheating on $a + 1$ questions strictly dominates cheating on a answers.

   (b) Which of the following gives a higher payoff to each student? (i) Every student cheats on $a + 1$ questions or (ii) every student cheats on a answers. Give the payoffs in each case.

3. **Iterated elimination of dominated strategies.** Use iterated elimination of dominated strategies to reduce the following games to smaller games in which iterated elimination of strictly dominated strategies cannot be used further. State the order in which you eliminate strategies. If you find a dominant strategy equilibrium, say what it is. Remember, you eliminate rows by comparing first entries in the two rows, and you eliminate columns by comparing second entries in the two columns.

   (a)

   $$\begin{pmatrix} & t_1 & t_2 & t_3 \\ \hline s_1 & (73, 25) & (57, 42) & (66, 32) \\ s_2 & (80, 26) & (35, 12) & (32, 54) \\ s_3 & (28, 27) & (63, 31) & (54, 29) \end{pmatrix}$$

[1]https://en.wikipedia.org/wiki/Tragedy_of_the_commons

(b)

$$\begin{pmatrix} & t_1 & t_2 & t_3 & t_4 & t_5 \\ \hline s_1 & (63,-1) & (28,-1) & (-2,0) & (-2,45) & (-3,19) \\ s_2 & (32,1) & (2,2) & (2,5) & (33,0) & (2,3) \\ s_3 & (56,2) & (100,-6) & (0,2) & (4,-1) & (0,4) \\ s_4 & (1,-33) & (-3,43) & (-1,39) & (1,-12) & (-1,17) \\ s_5 & (-22,0) & (1,-13) & (-2,90) & (-2,-57) & (-4,73) \end{pmatrix}.$$

4. Find all two by two matrix games that *do not* have a unique Nash Equilibrium.

5. Come up with another example of the tragedy of the commons. Now consider the Prisoner's Dilemma. Is this part of the tragedy of the commons family? What about the Stag Hunt?

6. Imagine a population of N stag hunters. Find the conditions on the distribution of the population that make the Nash equilibrium hunting stag the optimal strategy, and the conditions on the distribution of the population that make the Nash equilibrium hunting hare the optimal strategy.

CHAPTER 4

Mixed Strategy Nash Equilibria and Two-Player Zero-Sum Games

We now move on to a class of games in normal form where the game, or subgame after eliminating dominated strategies, has more than one strategy available to the players and those strategies are not dominated. Consider the payoff matrix given in Figure 4.1.

		Player 2	
		Move C	Move D
Player 1	Move A	(4,6)	(8,2)
	Move B	(7,3)	(1,8)

Figure 4.1: A GNF in which no strategy dominates another.

We can't use iterated elimination of dominated strategies here, since there are no dominated strategies to eliminate, and so we must come up with another solution. As before, some assumptions and definitions will help us start this process.

- Consider a GNF with players 1,...,n, with strategy sets $S_1, ..., S_n$, finite. Suppose each player has k strategies, denoted $s_1, s_2, ..., s_k$.

- We say player i has a *mixed strategy* if i uses s_i with a probability p_i. We refer to each $s_i \in S_i$ as a *pure strategy*.

- If the p_i associated with s_i is greater than zero, then we call s_i an *active pure strategy*.

- The mixed strategy of player i, denoted σ_i, is given by $\sigma_i = \sum_{i=1}^{k} p_i s_i$.

The following two points about playing a mixed strategy by either player will help us formulate what to do.

- If either player chooses a mixed strategy, it is because they are indifferent, or perhaps unsure, about which strategy they want to use. If they preferred one strategy over another

(Move A over Move B, or Move C over Move D, for example), then they would choose that strategy over another, rather than playing randomly.

- If we assume that the mixed strategy being played is part of the Nash equilibrium of the system, then the expected payoff to a given player from choosing one move over another must be equal.

Let's apply these ideas to the game given in Figure 4.1. For Player 1, who has the two strategies Move A and Move B, we can assign probabilities to those moves and generate an expected payoff. Let p_A be the probability that Player 1 uses Move A. This implies that Player 1 will use Move B with probability $1 - p_A$. That means the expected payoff to Player 2, given that Player 1 is indifferent about his moves, using Move C is

$$6p_A + (1 - p_A)3.$$

And if Player 2 uses Move D, then the expected payoff to Player 2 is

$$2p_A + (1 - p_A)8.$$

Since Player 2 knows that Player 1 is indifferent about what to do, the expected payoffs to Player 2 resulting from either of her moves is equal, and we can set the expected payoffs equal to one another and solve for p_A.

$$6p_A + (1 - p_A)3 = 2p_A + (1 - p_A)8$$

which gives $p_A = \frac{5}{9}$. If we use similar reasoning about how Player 1 knows that Player 2 is indifferent, then we can set up a similar equality of expected payoffs, given that Player 2 will use Move C with probability p_C and Move D with probability $p_D = 1 - p_C$. The resulting equations are

$$4p_C - (1 - p_C)8 = 7p_C + (1 - p_C)1,$$

solving this for $p_C = 0.7$ If both players are rational, then these probability values represent the best response of both players, and the solution is a Nash equilibrium. We can represent the mixed strategies as $\sigma_1 = \frac{5}{9}A + \frac{1}{3}B$ and $\sigma_2 = 0.7C + 0.3D$, and we often will write a *mixed strategy profile* as (σ_1, σ_2). In this case our mixed strategy profile is $(\frac{5}{9}A + \frac{4}{9}B, 0.7C + 0.3D)$, which also happens to be our mixed strategy Nash equilibrium.

4.0.1   THE FUNDAMENTAL THEOREM OF NASH EQUILIBRIA

Although Nash guaranteed that there were Nash equilibria for any game under certain conditions, his theorem does not help us find them for any given game. The following theorem helps us identify Nash equilibria.

Theorem 4.1   *The Fundamental Theorem of Nash Equilibria*
The mixed strategy profile $(\sigma_1, \sigma_2, ..., \sigma_n)$ is a mixed strategy Nash equilibrium if and only if:

1. *Given two active strategies of player i, s_i and s_j, the expected payoff to player i is the same regardless of whatever mixed strategies the other players employ.*

2. *The pure strategy s_i is active in the mixed strategy σ_i for player i, and the pure strategy s_j is not active in the mixed strategy σ_i, then $\pi_i(s_i) \geq \pi_i(s_j)$, the payoff to i when she uses s_i is greater than or equal to her payoff when she uses s_j, regardless of whatever mixed strategy the other players employ.*

The proof of this theorem can be found in [11]. What to take away from this theorem? Each player's active strategies are all best responses to the profile of the other player's mixed strategies, where "best response" means best response among pure strategies.

4.1   AN EXAMPLE WITH A 3-BY-3 PAYOFF MATRIX

Consider the following matrix:

	D	E	F
A	(9,-4)	(4,-2)	(-2,1)
B	(-2,4)	(5,3)	(0,2)
C	(5,2)	(2,4)	(1,5)

Notice that this matrix has no dominant strategies for either player. We can use the theory of mixed strategy Nash equilibria to find a MSNE where each player uses all three of their moves at least some of the time. We begin by finding the expected payoffs for player 1 for moves A, B, and C. Let p_D, p_E, p_F be the probability Player 2 uses move D, E, F, respectively:

$$\pi_1(A, \sigma_2) = 9p_D + 4p_E - 2p_F$$
$$\pi_1(B, \sigma_2) = -2p_D + 5p_E + 0p_F$$
$$\pi_1(C, \sigma_2) = 5p_D + 2p_E + 1p_F.$$

Set $\pi_1(A, \sigma_2) = \pi_1(C, \sigma_2)$ and $\pi_1(B, \sigma_2) = \pi_1(C, \sigma_2)$, giving us the two equations:

$$4p_D + 2p_E - 3p_F = 0$$
$$-7p_D + 3p_E - p_F = 0$$

plus the last equation where $p_D + p_E + p_F = 1$, since the probabilites associated with all of the moves must sum to 1. We now have a 3-by-3 linear system, which can be solved in a variety of ways. Use whichever technique you are most comfortable with, but here is how I would do it, using row reduction:

$$\left(\begin{array}{ccc|c} 4 & 2 & -3 & 0 \\ -7 & 3 & -1 & 0 \\ 1 & 1 & 1 & 1 \end{array} \right) \rightarrow \left(\begin{array}{ccc|c} 4 & 2 & -3 & 0 \\ 0 & 26 & -25 & 0 \\ 0 & -2 & -7 & -4 \end{array} \right) \rightarrow \left(\begin{array}{ccc|c} 4 & -2 & 3 & 0 \\ 0 & 26 & -25 & 0 \\ 0 & 0 & -116 & -52 \end{array} \right).$$

If you have forgotten how to row reduce, http://www.math.tamu.edu/~fnarc/psfil es/rowred2012.pdf can help. From here we can solve for the probabilites: $p_D = \frac{7}{58}$, $p_E = \frac{25}{58}$, $p_F = \frac{13}{29}$.

Now we can use a similar process involving Player 2's expected payoffs. Let p_A, p_B, p_C be the probability Player 2 uses move A, B, C, respectively.

$$\pi_2(\sigma_1, D) = -4p_A + 4p_B + 2p_C$$
$$\pi_2(\sigma_1, E) = -2p_A + 3p_B + 4p_C$$
$$\pi_2(\sigma_1, F) = p_A + 2p_B + 5p_C.$$

Set $\pi_2(\sigma_1, D) = \pi_2(\sigma_1, F)$ and $\pi_2(\sigma_1, E) = \pi_2(\sigma_1, F)$, giving us the two equations:

$$5p_A - 2p_B + 3p_C = 0$$
$$3p_A - p_B + p_C = 0$$

plus the last equation where $p_A + p_B + p_C = 1$, since the probabilites associated with all of the moves must sum to 1. Same deal as last time for the solution:

$$\begin{pmatrix} 5 & -2 & 3 & | & 0 \\ 3 & -1 & 1 & | & 0 \\ 1 & 1 & 1 & | & 1 \end{pmatrix} \rightarrow \begin{pmatrix} 5 & -2 & 3 & | & 0 \\ 0 & -1 & 4 & | & 0 \\ 0 & -7 & -2 & | & -5 \end{pmatrix} \rightarrow \begin{pmatrix} 5 & -2 & 3 & | & 0 \\ 0 & -1 & 4 & | & 0 \\ 0 & 0 & 30 & | & 5 \end{pmatrix}.$$

From here we can solve for the probabilites: $p_A = \frac{1}{6}$, $p_B = \frac{2}{3}$, $p_C = \frac{1}{6}$.

Therefore, our MSNE is $(\sigma_1, \sigma_2) = (\frac{1}{6}A + \frac{2}{3}B + \frac{1}{6}C, \frac{7}{58}D + \frac{25}{58}E + \frac{13}{29}F)$.

From this point we can calculate the expected payoffs for both players using the mixed strategy profile at the NE.

Calculating Expected Payoff

Basically, we sum the probability of every possible combination multiplied by its payoff. For Player 1, this looks like:

$$\pi_1(\sigma_1, \sigma_2) = p_A p_D \pi_1(A, D) + p_A p_E \pi_1(A, E) + p_A p_F \pi_1(A, F) + \cdots + p_C p_F \pi_1(C, F)$$
$$= (\frac{1}{6})(\frac{7}{58})9 + (\frac{1}{6})(\frac{25}{58})4 + (\frac{1}{6})(\frac{13}{29})(-2) + \cdots (\frac{1}{6})(\frac{13}{29})1 = 1.913793103.$$

For Player 2, this looks like:

$$\pi_2(\sigma_1, \sigma_2) = p_D p_A \pi_2(A, D) + p_D p_B \pi_2(B, D) + p_D p_C \pi_2(C, D) + \cdots + p_F p_C \pi_2(C, F)$$
$$= (\frac{1}{6})(\frac{7}{58})(-4) + (\frac{1}{6})(\frac{25}{58})4 + (\frac{1}{6})(\frac{13}{29})(2) + \cdots + (\frac{1}{6})(\frac{13}{29})5 = 2.333333333.$$

Since $\pi_1(\sigma_1, \sigma_2) < \pi_2(\sigma_1, \sigma_2)$, we would say this game favors Player 2.

4.2   TWO-PLAYER ZERO SUM GAMES

A zero sum game is a game, usually in normal form but not always, that forces the total sum of the payoffs to be zero. What one player wins, the other has to lose. We will restrict our attention to the finite game. Two-player zero sum games (TPZSG), also called **matrix games** have some interesting properties, but perhaps the most significant is the *Minimax Theorem.*

Theorem 4.2   The Minimax Theorem.   *For every finite two–player zero sum game:*

- *there is some value V, called the **game value**;*

- *there exists a mixed strategy for Player 1 that guarantees an average payoff of at least V to Player 1, regardless of Player 2's strategy;*

- *there exists a mixed strategy for Player 2 that guarantees an average payoff of at most –V (a loss) to Player 2, regardless of Player 1's strategy; and*

- *if both players use the strategies that guarantee their average payoffs, the mixed strategy profile is a Nash equilibrium.*

A proof of this theorem is provided at `http://www.math.udel.edu/~angell/minimax` but it shouldn't be too hard to see why it is true. If V is zero, we define the game as *fair.* If V is not equal to zero, we define the game as *unfair.* If V is positive, the game is weighted in Player 1's favor, and if V is negative, the game is weighted in Player 2's favor.

4.2.1   THE GAME OF ODDS AND EVENS

Players 1 and 2 simultaneously call out the numbers 1 or 2. If the sum is odd, Player 1 wins the sum of the numbers called out. If the sum is even, Player 2 wins the sum of the numbers called out. The loser must pay the winner the sum. The payoff matrix for this game is given in Figure 4.2.

		Player 2	
		1	2
Player 1	1	(-2,2)	(3,-3)
	2	(3,-3)	(-4,4)

Figure 4.2: The Odds and Evens Game payoff matrix.

Which player does this game favor? It may not be obvious from the start, but we can explicitly calculate the mixed strategy Nash equilibrium for this game, and show it is also the Minimax strategy for both players. This will also tell us which player the game favors.

- Let p be the probability that Player 1 uses the strategy 1, and $1 - p$ for 2.

- Let q be the probability that Player 2 uses the strategy 1, and $1 - q$ for 2.

- If we use the method defined earlier, multiplying probabilities and payoffs, setting equations equal to one another and solving, we end up with $p = \frac{7}{12}$ and $q = \frac{7}{12}$.

If we calculate the average payoffs to Player 1, we see that $\pi_1 = (\frac{7}{12})^2(-2) + 2(\frac{7}{12})(\frac{5}{12})(3) + (\frac{5}{12})^2(-4) = \frac{1}{12}$, and $\pi_2 = (\frac{7}{12})^2(2) + 2(\frac{7}{12})(\frac{5}{12})(-3) + (\frac{5}{12})^2(4) = \frac{-1}{12}$. These payoff values are what we defined as the game value V. These average payoffs indicate that this game played at the Minimax strategy favors Player 1, and by the Fundamental Theorem of Nash Equilibria, this Minimax strategy is also the NE.

4.3   DOMINATION OF TWO-PLAYER ZERO SUM GAMES

When representing the payoffs to the players in a matrix game, we will now use the convention that only Player 1's payoffs will appear in the matrix. This is due to the fact that the payoffs are symmetric, and if we wanted to display Player 2's payoff we would just multiply every value in the matrix by -1. The Odds and Evens game would be represented as

		Player 2	
		1	2
Player 1	1	-2	3
	2	3	-4

4.3.1   SADDLE POINTS

A saddle point, reminiscent of the saddle points in multidimensional calculus, of a matrix game is the term in the matrix that is the minimum of its row and the maximum of its column. An example is given in Figure 4.3.

$$A = \begin{pmatrix} -1 & 3 & 4 \\ 2 & 4 & 6 \\ 1 & 0 & 2 \end{pmatrix}$$

Figure 4.3: The saddle point of this matrix is at the term a_{21}.

The value at a_{21} is 2, which is the minimum of its row and the maximum of its column. Recall that the payoffs are with respect to Player 1.

4.3.2   SOLVING TWO-BY-TWO GAMES

Consider the following two-by-two matrix

$$B = \begin{pmatrix} a & b \\ c & d \end{pmatrix}.$$

Strategy for finding the solution.

- Check for a saddle point. If there is a saddle point, that value is the game value V.

- If there is no saddle point, we can find the mixed strategy Nash equilibrium by finding the minimax strategy.

If there is no saddle point, we can characterize the solutions using the following equation:

$$pa + (1 - p)c = pb + (1 - p)d \qquad (4.1)$$
$$(a - b)p = (1 - p)(d - c) \qquad (4.2)$$
$$p = \frac{d - c}{(a - b) + (d - c)}. \qquad (4.3)$$

The average return to Player 1 based on this probability is

$$V = \pi_1 = pa + (1 - p)b = \frac{(ad - bc)}{a - b + d - c}.$$

We can use a similar technique to show that V is the same for Player 2 and that

$$q = \frac{d - b}{a - b + d - c}.$$

Note: Often in TPZSG, it is necessary to reduce the game to a smaller matrix. If it happens to reduce down to a two by two matrix, then the solution to the larger game can be characterized by the solution given.

For example, if we consider the matrix game given in

$$G = \begin{pmatrix} 1 & 4 & 10 \\ 2 & 5 & 1 \\ 3 & 6 & 9 \end{pmatrix}.$$

Column 2 is dominated by column 1, and row 1 is dominated by row 3, so this game is now reduced to

$$H = \begin{pmatrix} 1 & 10 \\ 3 & 9 \end{pmatrix}.$$

The new matrix H has a saddle point, at a_{21}, with a value of 3. If you attempted to solve this matrix using the characterization given, you would get a probability that exceeded 1 or was less than zero.

4.4   GOOFSPIEL

Goofspiel [13] is a two-player zero sum game that defies complete conventional analysis. It is played with a standard 52 card deck and the rules are as follows.

- Remove 1 suit (13 cards) from the deck. This is usually clubs, but it is irrelevant which suit you remove.

- Player 1 receives the 13 cards of the hearts suit, and Player 2 receives the 13 cards of the diamonds suit.

- The 13 remaining spades are shuffled and placed in between the two players.

- One is turned face up. The two players simultaneously choose a card and discard it face up. Whichever player discarded the card with the highest value (Ace being worth 1, King is worth 13) wins the spade card that is turned up in the middle.

- The worth of the spade card is added to the player's score, and subtracted from the other player's score (it is possible to have negative score).

- If both players discard a card of equal value (not necessarily to the spade), then they receive nothing and the spade card is lost.

- Repeat until no cards are left. The losing player must pay their score to the winner (perhaps in dollars).

Interestingly, in work done by Sheldon in 1971, it is shown that the best course of play against an opponent that plays randomly is to match your bid to the value of the spade card. That is the only result currently known about how to play this variation of the game. Another variation is when the card in the middle is hidden, and then it is best to just play randomly.

Goofspiel has recently been featured in a round robin tournament [14] that further shows that there is a great deal of analysis still required to get a handle on this game, if it is at all possible. At this point, we still don't really know how to play Goofspiel.

4.5   EXERCISES

1. Reduce the following game G and find the MSNE of the subgame H. Find the expected payoff to both players.
$$G = \begin{pmatrix} (4,1) & (5,3) & (6,4) \\ (8,1) & (9,5) & (1,2) \\ (2,0) & (4,1) & (0,2) \end{pmatrix}.$$

2. If G represents a TPZSG, find its game value if

$$G = \begin{pmatrix} 2 & 5 & 8 \\ 1 & 4 & 1 \\ 3 & 4 & 5 \end{pmatrix}.$$

3. (Short Essay) Is the two player zero sum game model appropriate for war between nations? Be sure to think about what is involved in the fighting and the resolution of a war. World War 1 and World War 2 are fine examples, but there have been wars of all sizes throughout the course of recorded human history. Do some research on a few wars and give some examples about why a two player zero sum game model is or is not a good model.

4. Imagine you are playing Goofspiel and you know ahead of time that your opponent is using the strategy: *Match the value of the upturned card.* Find the strategy that maximizes your score every round, and prove it is a maximum score. Now generalize this strategy for any finitely sized set of Goofspiel, and show that your strategy still maximizes your score every round.

5. Come up with a strategy for an agent to play Goofspiel in a tournament, where you are uncertain of what other strategies will be submitted. You can write your strategy in English, or submit a finite state machine.

CHAPTER 5

Mathematical Games

5.1 INTRODUCTION

We now change our focus from classical game theory to the theory of mathematical games. Mathematical game theory covers a broad spectrum of games with relatively simple rules and constraints, but usually reveals deep mathematical structure. There are two books, *Winning Ways for Your Mathematical Plays* by Berlekamp, Conway and Guy written in 1982, and *Fair Game*, by Guy in 1989, that offer a wide variety of interesting mathematical games and analyses. This chapter will focus on mathematical games with perfect information. There are many other mathematical games with imperfect information that are just as interesting, but far more complex to analyze.

5.1.1 THE SUBTRACTION GAME

The subtraction game, also known as the takeaway game, invented by Thomas S. Ferguson, is a game with two players.

The Subtraction Game

- There are two players. Player 1 makes the first move, and Player 2 makes the second move.

- There is a pile of 10 tokens between the players.

- A move consists of taking 1, 2, or 3 tokens from the pile.

- The player that takes the last token wins.

Let's define the **game state** of the subtraction game as how many tokens are still left in the pile, and which player is going to make a move. We use an ordered pair (Game state, Player) to tell us about the game state. With this definition we can use a technique similar to backward induction to show us the solution to this game.

Assume the game state has changed from (10,1) to (4,2). This means that there are 4 tokens left in the pile, and it is Player 2's turn. If Player 2 takes 1, 2, or 3 tokens (all of her available moves), then there remains 3, 2, or 1 tokens, respectively, in the pile. All of the game states (3,1), (2,1), and (1,1) are what are called **winning states** for Player 1. This implies that (4,2) was a **losing state** for Player 2. If the game state was (4,1), that is a losing state for Player 1 and a winning state for Player 2. We can continue with this style of analysis all the way up to 10 tokens, and we can see that the game states $(4,i)$, $(8,i)$ are losing states for Player i, $i = 1, 2$.

Let's focus on the winning states for Player 1 in the 10 token subtraction game. For simplicity, the notation X will be used to denote the game state (X,1). Consider Figure 5.1. The winning and losing states for Player 1 are given.

Tokens	0	1	2	3	4	5	6	7	8	9	10
	L	W	W	W	L	W	W	W	L	W	W

Figure 5.1: The winning and losing states for Player 1 in the subtraction game.

Keep in mind that Figure 5.1 is in reference to Player 1. If there are 0 tokens, Player 1 loses. If Player 1 has their turn when the game state has 4 or 8 tokens, Player 1 will lose if Player 2 uses the correct strategy. In all other states, Player 1 can maneuver the game into a losing state for Player 2.

The generalized subtraction game is defined in the following manner. Notice that a repeating pattern forms of length 4, LWWW. If we extended our result to more than ten tokens, it should not be too difficult to see why this pattern extends for any number of tokens. The size of the repeating pattern becomes relevant in the general game.

- Let $\mathbb{S}$ be a set of positive integers, called the *subtraction set*.

- Let there be n tokens, and two players, with the convention that Player 1 moves first.

- A move consists of removing $s \in \mathbb{S}$ tokens from the pile.

- The last person to move wins.

There is a known result that shows that subtraction games with finite subtraction sets have eventually periodic win-loss patterns [15]. This repeating pattern has a special name, called the **Nim sequence**.[1] There are backward induction style algorithms that will find the Nim sequence under certain conditions, but we don't need them for our purposes. For example, with the subtraction set $\mathbb{S} = \{1, 3, 4\}$, the Nim sequence is of length seven and is LWLWWWW. The Nim sequence and its length are of importance in the following conjecture.

Conjecture. Consider a set $\mathbb{S}$ in the subtraction game with finite tokens, n. Use $\mathbb{S}$ to construct the Nim sequence of W's and L's of length l. Then the number of tokens $n \bmod l$ determines which player will win with optimal play. If $n \bmod l = k$, and k is a winning position, then Player 1 will win the game with n tokens with optimal play. If k is a losing position, then Player 1 will lose if Player 2 plays optimally.

If we consider the game with $\mathbb{S} = \{1, 2, 3\}$, that has a Nim sequence of length 4. If the game originally starts with 2,001 tokens, we do not need to build the pattern out to that point.

---

[1] There are other definitions given for *Nim sequence* in the combinatorial games literature. The one I am using is from a recent paper.

Instead, if we find 2001%4 = 1, and 1 is a W position for Player 1, then we know that Player 1 can win the game when it starts with 2,001 tokens.

While there is an algorithm for determining the Nim sequence, there is still the following open problem: Is there a way to determine the Nim sequence from a subtraction set directly?

5.2 NIM

Nim, coined by Charles L. Bouton in 1901, is an extension of the subtraction game. Rules of Nim:

- There are two players, Player 1 moves first.

- There are now n piles of tokens, rather than one. The piles do not necessarily have the same number of tokens, but they all have at least one.

- A move consists of taking a positive number of tokens from a single pile.

- The winner is the player who takes the last token.

We can do some similar analysis to what we did in the subtraction game. We can represent the game state of Nim using an ordered $(n + 1) - tuplet$. The first n terms in the game state represent the amount of tokens in each pile, and the last term indicates whose turn it is. If we consider the 3-pile game with n_1, n_2, n_3 tokens and it is Player 2's move, then the game state would be written as $(n_1, n_2, n_3, 2)$. Similar to what was done in the subtraction game, we will evaluate all positions relative to Player 1 rather than Player 2.

We'll restrict our analysis to the 3-pile game, but these results can easily be extended to the n-pile game. First, consider the game state where only one pile has any remaining tokens. This is clearly a winning state, since Player 1 can simply remove the whole pile.

Two non-empty piles have two possibilities. Either the piles have an equal number of tokens, or not. If the piles are equal, that is a losing state. Can you think of why that should be so? If the piles are not equal, that is a winning state, since the player making the move can now make the piles equal.

Three non-empty piles are a good deal more interesting. Consider (1,1,2), (1,1,3), and (1,2,2). Are these winning or losing states for Player 1? It shouldn't take too long to figure out that (1,1,2) is a winning state, (1,1,3) is also a winning position, and (1,2,2) is also a winning state. Rather than working it out for every single case, there is an easier way to determine the winning and losing states in Nim. We need something called the Nim-sum, which is reminiscent of the Nim sequence, but unrelated.

Definition 5.1 The Nim-sum of two non-negative integers is their addition without carry in base 2.

If you need a refresher on how to turn base 10 numbers into their base 2 representations, `http://www.wikihow.com/Convert-from-Decimal-to-Binary` offers an excellent tutorial.

Calculating the Nim-sum

The following example is meant to clarify how find the Nim-sum of a set of positive integers. Consider 14, 22, and 31 as the numbers whose Nim-sum we would like to find. First, convert all of the numbers into base 2. $14 = (01110)_2$, $22 = (10110)_2$, and $31 = (11111)_2$. Stack the base 2 expansions (order is irrelevant here, since addition is commutative) and add them columnwise, without carry. So,

$$
\begin{array}{cccccc}
  & 0 & 1 & 1 & 1 & 0 \\
  & 1 & 0 & 1 & 1 & 0 \\
+ & 1 & 1 & 1 & 1 & 1 \\
\hline
  & 0 & 0 & 1 & 1 & 1 \\
\end{array}
$$

Notice that in the middle column three ones are added, but because we are working in base 2 we need to use modular arithmetic, $(1 + 1 + 1) \mod 2 = 1$. The Nim-sum of 14, 22, and 31 yields $(00111)_2$, or 7 in base 10.

Theorem 5.2   *A game state $(x_1, x_2, ..., x_n)$, where $x_i \geq 0, i = 1, 2, ..., n$, is a losing state in Nim if and only if the Nim-sum of its piles is 0.*

The nice thing about this theorem is that is works for an arbitrary finite number of piles. If the Nim-sum does not produce 0, then those piles represent a winning position for the player that makes an optimal move. The proof of this theorem, given by C. L. Bouton in 1902, contains the algorithm that shows you what an optimal move will be, and for the interested reader, that proof can be found here: `http://www.stat.berkeley.edu/~peres/gtlect.pdf`. Without going into details about the proof, we offer the algorithm below.

Algorithm for Finding a Winning Move from a Winning Position

Form the Nim-sum column addition
Look at the left-most column with an odd number of 1's
Change a 1 in that column to a 0
Change the numbers in the row of the changed 1, such that there are an even number of 1's in each column

---

Note: There may be more than one winning move based on this algorithm.

5.2.1 MOORE'S NIM

There are several variations on the game of Nim, but we will take a look at this one. Moore's Nim, developed be E.H. Moore in 1910, also has the name Nim_k. There are n piles of tokens and the rules are the same as Nim, except that in each move a player may remove as many chips as desired from any k piles, where k is fixed and $k < n$. At least one chip must be removed from at least one pile. He also came up with the following theorem.

Theorem 5.3 $(x_1, x_2, ..., x_n)$ *is a losing position in Nim_k if and only if when the x_i's are expanded into base 2 and added in base $k + 1$ without carry, the sum is zero.*

There is an algorithm similar to Buton's that allows a player in a winning position to find the optimal move, the only difference being the number produced by the Nim-sum needs to be evaluated in base $k + 1$.

5.3   SPROUTS

Sprouts is a game invented by Conway and Paterson in the 1960's, featured in *Winning Ways for your mathematical plays* and despite its simplicity it caught the mathematical community by storm for a while. Sprouts is another excellent example of how a mathematical game with simple rules can have deep and interesting properties. Sprouts is played in the following manner.

- Sprouts is a two-player game, with the convention that Player 1 moves first.

- On a piece of paper, or a computer screen in this age, a finite number of dots are placed in distinct locations on the paper.

   Note: The location of the dots is irrelevant to actual game play, but our brains trick us into thinking that it matters.

- A move consists of drawing an arc that connects one dot to another, possibly itself. This arc may not pass through another dot on the way, and it may not cross itself. Once the arc is completed, a new dot is placed at some point on the arc connecting the dots, that is not either endpoint.

- You cannot draw an arc to or from a dot that has degree 3 (meaning there are three arcs already ending at the dot).

- An arc cannot cross another arc.

- The game ends when no more arcs can be drawn.

- The player who draws the last arc wins.

   Consider the following example game with three dots, shown in Figure 5.2.
   The first player draws an arc from the top dot to the right dot, shown in Figure 5.3. In the middle of this arc, the first player places a dot. This dot counts as already having two arcs sprouting from it.

Figure 5.2: The starting configuration for a three-dot game of Sprouts.

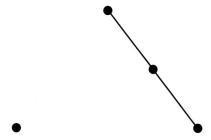

Figure 5.3: The first move for a three-dot game of Sprouts.

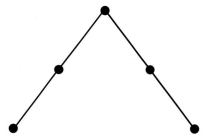

Figure 5.4: The second move for a three-dot game of Sprouts.

The second player then draws an arc from the top dot to the left dot and places a dot in the middle of that arc, shown in Figure 5.4. The first player then draws an arc from the left dot to the right dot and places a dot in the middle of that arc, shown in Figure 5.5.

The second player then draws an arc from the right dot to the dot in between the top and right dot, and places a dot in the middle of that arc, shown in Figure 5.6. Notice that the right dot and the dot in between the top and right dot now have circles drawn around them, to indicate they are have three arcs connected to them and can no longer be used in the game. The first player then draws an arc from the top dot to the left dot, and places a dot in the middle of

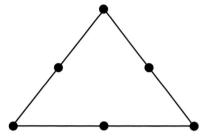

Figure 5.5: The third move for a three-dot game of Sprouts.

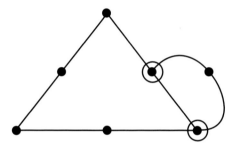

Figure 5.6: The fourth move for a three-dot game of Sprouts.

that arc, shown in Figure 5.7. Similar to the last picture, the top dot and the left dot now have circles drawn around them to indicate they can no longer be used in the game.

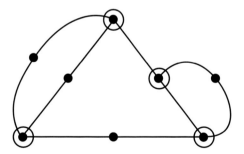

Figure 5.7: The fifth move for a three-dot game of Sprouts.

The second player then draws an arc from the middle dot of the first arc connecting the top and left dot, and the middle dot of the second arc connecting the top and left dot, while placing a dot in the middle of that new arc. The two dots used to start the arc have circles drawn around them.

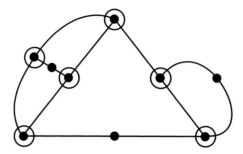

Figure 5.8: The six move for a three-dot game of Sprouts.

The first player then draws an arc from the right most dot to the bottom middle dot, and places a dot in the middle of that arc, shown in Figure 5.9. The game ends, as Player 2 cannot make a move, and Player 1 wins.

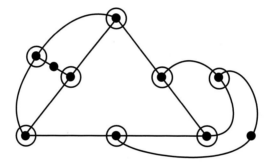

Figure 5.9: The final move for a three-dot game of Sprouts.

If a game of Sprouts starts with d dots, then the maximum number of moves that can occur is $3d - 1$. Each dot starts with 3 *lives* (how many curves can be connected to a vertex), and each move reduces 2 lives from the game and adds one new dot with one life. Thus, each move globally decreases the number of lives by 1, and at most the number of moves is $3d - 1$. The minimum number of moves is $2d$, and it depends on creating enclosed regions around dots with remaining lives. Despite the fact that a finite number of dots means there are a finite number of strategies, analysis of Sprouts has required the invention of new representations and some interesting mathematics involving *Nimbers*, also known as Sprague-Grundy numbers. Games that begin with 2–44 dots have been completely analyzed, it is known whether the first or second player will win for those games. The game that starts with 45, 48, 49, 50, 51, and 52 dots are currently unknown. The game with 53 dots is the highest known starting configuration for Sprouts. A list of who has done the analyzing and what game configurations are known can be found at http://sprouts.tuxfamily.org/wiki/doku.php?id=records.

The paper written by Julien Lemoine and Simon Viennot, "Computer analysis of sprouts with nimbers," is an excellent read on this topic, but it involves mathematics beyond the scope of this course. However, one interesting conjecture given by [16], is at least supported by their work:

The Sprouts Conjecture. The first player has a winning strategy in the d-dot game if and only if d modulo 6 is 3, 4, or 5.

This conjecture has been supported up to 44 dots, but as of this time there is no formal proof, and so it remains an open problem.

5.4   THE GRAPH DOMINATION GAME

The game of graph domination, invented by Daniel Ashlock in 2009, is played on a combinatorial graph. There is a similar game of graph domination involving the coloring of vertices and finding minimal spanning trees, but this game does not use those mechanics. The graph domination game is a game played between two players. Each player has a pile of tokens and by convention, one player is labeled Red, the other Blue, and Red moves first. Consider the graph in Figure 5.10, we call it a *game board*.

Each circle on the board, including the entry points, is called a *vertex*. The lines joining the vertices are called *edges*. Two vertices with an edge between them are *adjacent*. The rules of the Graph Domination Game are as follows.

1. Each player has an entry point. The red player's entry point is red, the blue player's entry point is blue.

2. On their turn a player may either place one of their tokens on their entry point, if it is currently unoccupied, or they may move one of their tokens from a vertex to an adjacent vertex, except as specified in rule 3.

3. No player may place or move a token into a vertex that is adjacent to a vertex occupied by one of their opponent's tokens.

4. A player must place or move a token on each turn unless this is impossible because of rule 3. In this case the player does nothing during that turn.

5. The game continues until neither player can move.

6. At the end of the game a player's score is three point for each vertex occupied by one of their tokens. Each unoccupied vertex is worth one point to a player for each token he has adjacent to that vertex.

7. The player with the highest score is the winner, if the scores are equal the game is a tie.

8. Players may wish to play two games with each player going first in one of the games and add the scores of the two games.

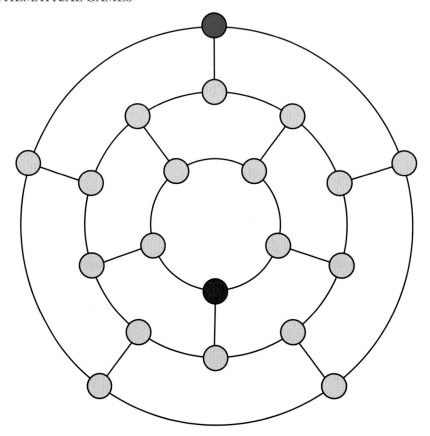

Figure 5.10: Game board.

Four examples of different game boards are presented in Figure 5.11.

Let us define a *ring* as a group of vertices that rest on a given circle created by edge arcs when the game board is drawn. The *Red Ring* is the ring on which the red entry point is located, and the *Blue Ring* is the ring on which the blue entry point is located.

This game has yet to be analyzed in any real depth. There are still many open questions.

- Does the first mover have the advantage?

- All of the game boards are symmetric, and more importantly, vertex transitive. Is there an optimal number of vertices per ring that promotes game complexity while keeping it interesting?

- How much does edge structure affect the game? The current game boards are all 3-regular (meaning three edges are connected to every vertex). What do 4-regular game boards look

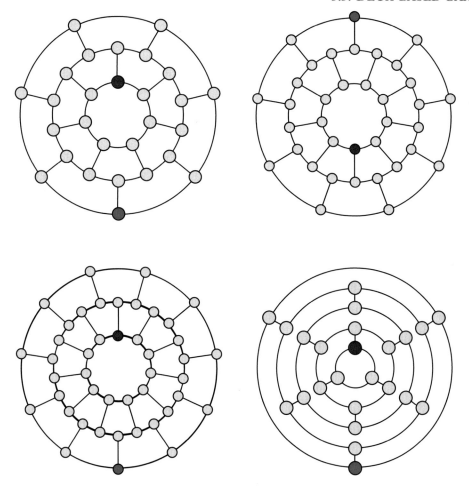

Figure 5.11: Four examples of different game boards.

like? What strategy should be employed there? Does that strategy deviate signficantly from the 3-regular board strategy?

- What affect does the number of rings between the Blue and Red ring have on the strategy employed?

5.5   DECK-BASED GAMES

Deck-based games were invented by asking the question: "What if, instead of being able to move any way we wanted, we played prisoner's dilemma with a deck of cards printed with cooperate and defect?" The answer to this question turned out to be much deeper and more startling than expected. Formally, a *deck-based game* is created when a restriction is placed on the available moves in a typical mathematical game, such as rock-paper-scissors, or the Prisoner's Dilemma, [17]. This restriction can come in the form of limiting the number of times a player can use a certain move in a given game, or it can be in the form of a player having access to a subset of the available moves. If the moves are placed on playing cards, then a new kind of game is created, called a deck-based game.

General Mechanics of Deck-Based Games

A deck-based game is called *strict* if a player must use all of the cards in his hand. The original version of a mathematical game, such as the Subtraction Game, is called the *open* version of the game. There exists a transition from deck-based to open games by allowing the total number each move available to a player to grow larger than the number of rounds played. A deck-based game can be derived from games that have sequential or simultaneous play, but we'll focus on games with simultaneous play in this book. From now on we'll refer to moves as cards, encapsulating the fact that there are restrictions on both how many times a move may be used, and how many moves are available. A few more definitions.

- A *hand* is the set of cards currently available to a player.

- The *deck* is the total set of cards available in the game.

- A deck can be *shared*, in games such as Poker or Blackjack.

- A deck can be *individual*, which is a set of cards that one player has access to, such as in games like Magic: The Gathering™.

- If the players are choosing cards from a limited subset of the deck, this is called *card drafting*.

- If the players are constructing their individual decks as play proceeds, usually though card drafting, this is called *deck building*.

5.5.1 DECK-BASED PRISONER'S DILEMMA

As an example of how imposing restrictions on a classic mathematical game fundamentally changes its nature, we now consider the deck-based Prisoner's Dilemma (DBPD). Let the cards be labeled with **C** or **D**, the moves cooperate and defect, respectively. If we limit the game to one round and one of each type of card, we've effectively recreated the original Prisoner's Dilemma game.

If instead we increased the number of rounds and the amount of each type of card available to a player, a novel development occurs. Adding restrictions to the Prisoner's Dilemma actually creates three new games, dependent on the payoff matrix. Recall the general payoff matrix for the Prisoner's Dilemma.

	Player 2	
	C	D
Player 1 C	(C,C)	(S,T)
D	(T,S)	(D,D)

$$S \leq D \leq C \leq T, \; 2C \geq S + T$$

Figure 5.12: The generalized payoff matrix for the Prisoner's Dilemma.

Let's begin our analysis with the following assumptions. Assume the game lasts for N rounds, finite, and that each player has sufficient cards in their hand to play out all N rounds, satisfying the strict condition. If we assume that both players have the exact same distribution of cards in their hand at the start of the game, an unexpected result happens.

Theorem 5.4   *In the strict DBPD, if both players have the exact same distribution of cards in their hand at the start of the game and do not add new cards to their hand, then their score will be equal regardless of what strategy they employ.*

Proof. It is not too hard to see the case with one card is true. If we extend to the case where k cards are in play, that means that each player has some number d of defect cards and some number c of cooperate cards, such that $c + d = k$. Every time a player discards a card, either the card is matched, meaning D with D or C with C, or not matched, C with D or D with C. If there is a match, both players have $c - 1$ or $d - 1$ of cooperates or defects left, and they both have the same amount of each type of card. If there is a match, both players receive the same payoff of C or D. If there is no match, then we get the following. Without loss of generality, if Player 1 defects and Player 2 cooperates, Player 1 receives a payoff of T and Player 2 receives a payoff of S. At some round later on, Player 2 will have one more defect card than Player 1, and Player 1 will have one more cooperate card than Player 2. When those cards are subsequently played, perhaps during the last round of the game, the payoff to Player 1 will be S and the payoff to Player 2 will be T, thus balancing out the difference in scores. By the end of the game, both players will have the same score. □

A natural question to ask based on this is, "If both players receive the same score, what is the point in playing the game?" If there are just two players playing the game, then the answer is "There is no point, we should not play this game." However, consider the case where there are P players playing in a round robin tournament. Let your score during the tournament be the sum of the scores from every game you play. Theorem 5.4 indicates that you will achieve the same score as your partner during a single game with N rounds, but it says nothing about what the maximum or minimum score of that game will be.

What your maximum or minimum score per game will be depends on the payoff matrix. If $T + S > C + D$, a player attains the maximum score if they discoordinate with their opponent, playing D's against C's, and C's against D's. Coordination with your opponent will produce the minimum score. If $T + S < C + D$, then coordinating will produce the maximum score, and discoordination will produce the minimum score. If $T + S = C + D$, then the game is trivial and the maximum and minimum score will be equal, and all players will receive the same score each game.

An open question that has been partially answered by [18] is, "What is the optimal strategy to employ when the game is no longer strict?" meaning that there are more cards in a hand than number of rounds.

5.5.2   DECK-BASED ROCK-PAPER-SCISSORS(-LIZARD-SPOCK)

In the classic game Rock-Paper-Scissors (RPS), two players simulatenously choose a move from the set $\{Rock, Paper, Scissors\}$. We restrict our attention to the zero sum version of RPS. The payoff matrix, with respect to Player 1, for RPS is given below:

		Player 2		
		Rock	Paper	Scissors
Player 1	Rock	0	-1	1
	Paper	1	0	-1
	Scissors	-1	1	0

For the purpose of analyzing the available moves in a deck-based game, it is useful to employ a directed graph representation, as in Figure 5.13. We refer to directed graphs as *digraphs*.

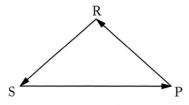

Figure 5.13: The directed graph representation of Rock-Paper-Scissors. R for Rock, P for Paper, S for scissors.

The digraph almost instantly conveys exactly how RPS works without the need for a payoff matrix. When using a digraph representation, specifying the payoff to the winner and loser is necessary and can be done using a typical payoff matrix, or assigning a pair of weights to each edge in the digraph. Notice that there are no arrows drawn from a move to itself; this is because a move cannot win or lose against itself, in this particular game. Since each edge is worth the same amount we leave the weighting out of the diagram. The convention for most games is that the payoffs for Losing, Tieing, and Winning are $L \leq T \leq W$, and at least one of those inequalities should be strict for a game to be at least mildly interesting. We can use a digraph for more complex games, like Rock-Paper-Scissors-Lizard-Spock (RPSLK). Consider the digraph in Figure 5.14.

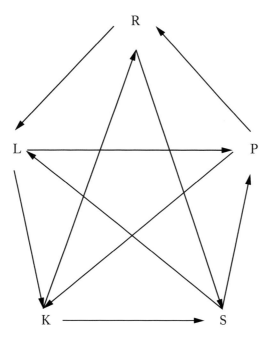

Figure 5.14: The directed graph representation of Rock-Paper-Scissors-Lizard-Spock. L is for Lizard, and K is for Spock.

The digraph representation also allows us to analyze a game rapidly. In both the RPS and RPSLK digraphs, it quickly becomes clear that no move is dominant, and we can formalize this with the concept of a *local balance*. The *outdegree* of a vertex is the number of arrows leaving that vertex, which is the number of moves that move defeats. The *indegree* of a vertex is the number of arrows incoming to that vertex, which is the number of moves that defeat that move. The local balance is the ratio of a vertex's outdegree over its indegree. Notice that in both RPS and RPSLK, the local balance of each vertex is 1. We call a game *balanced* if this is the case. Not every game can, or should, be balanced. We consider Rock-Paper-Scissors-Dynamite (RPSD), where Dynamite can beat all other moves and is not defeated by any move. RPSD is pictured in Figure 5.15.

The local balance of the Dynamite move is ∞, and the local balance of R, P, and S is $\frac{1}{2} = 0.5$. Some things we can conclude from using the digraph's local balance to analyze a game.

- Moves with a local balance of ∞ are unbeatable. If you are designing a deck-based game, include moves like this sparingly, or not at all.

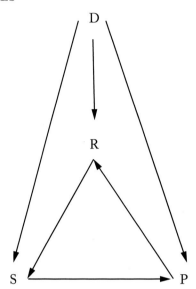

Figure 5.15: The directed graph representation of Rock-Paper-Scissors-Dynamite. D is for Dynamite.

- Moves with a balance of 0 (where they have some indegree > 0 and an outdegree of 0) are worthless moves and should never be included in a game, unless there is some extra point to consider.

- Moves with a balance of one, or close to it, should make up the bulk of the deck in a game.

- It is important to consider the cycles created by a deck-based game. In RPS, there is only one Hamilton cycle of edge length 3 in one direction, but there are two Hamilton cycles of edge length 5 in RPSLK. The relationship of cycle length to number of players in a game should be considered.

5.5.3 DECK-BASED DIVIDE THE DOLLAR

Deck-based Divide the Dollar (DBDD) is a card game based off of Divide the dollar and is another example of taking a classic mathematical game and applying restrictions to it.

Divide the Dollar

Divide the dollar is a simplified version of Nash's Bargaining Problem [19]. One dollar sits on the table between two players, and the players simultaneously make demands on how much of the dollar they want. If the sum of their demands are equal to or less than a dollar, they both receive their demand. If the sum of their demands exceeds a dollar, they both receive nothing.

Divide the dollar can be generalized in a number of ways. Modifying the goal amount is one, and an infinite family of generalizations of divide the dollar to any number of players appears in [20].

How to play DBDD

Deck-based Divide the Dollar (DBDD) has the following rules.

- The game is played by N players.

- There is a goal deck that contains a distribution of goal amounts placed on cards. This distribution is dependent on how many players are involved in the game, and on the possible distribution of the players' hands.

- Each player has a deck to draw from, and draws some number of cards from their deck at the beginning of the game. When a player discards, they draw a new one from their deck.

- During a given round, each player discards a card. If the sum of the played cards is less than or equal to the goal card, each player receives a payoff equal to the value of the card they played. If the sum exceeds the goal card, each player receives a payoff of zero.

- The player with the highest score at the end of the game is the winner.

- Game play can be simultaneous, or sequential, with the player who goes first rotating around the player group.

DBDD has novel dynamics to consider. In the open game of Divide the dollar, where a player may bid any value, the Nash equilibrium is that all players should bid the goal divided by the number of players, rounding down if the amounts are discretized. That is no longer the case, since the moves are restricted by the cards available in a player's hand. If a player has an idea of the distribution of goal cards and the distribution of cards in the other players' hands, they may be able to make an informed decision about optimal strategy. Sequential vs. simultaneous play also drastically changes the game and how strategies are implemented. It is clear that the player who plays the last card has tremendous advantage over the player who plays the first card. Effectively, they often will control whether or not the goal card will be exceeded. However, that analysis still needs to be completed at this point.

5.5.4   FLUXX-LIKE GAME MECHANICS

Fluxx, created by Looney Labs, is a very entertaining card game. The cards that are played modify the conditions under which the game can be won, as the game is played. So the payoff for any given card is determined by in game conditions that change as the game is played. Deck-based games are interesting in and of themselves, but adding a Fluxx inspired game mechanic can move the deck-based game in a very different direction than the original open version. If we allow certain cards to change the payoff of moves, then we are broadening the possibilities of events that can occur in the game.

Fluxx-like Deck-based Prisoner's Dilemma

At the risk of our acronyms getting out of hand, FDBPD has Fluxx-like cards as part of the deck from which players are allowed to draw. There are many possible choices to make about what sort of cards can change the nature of the game, but we can keep it simple and show the complexity that can fall out of an adjustment like this one. Assume there is a new kind of card available to each player, called a *force* card. This card is played simultaneously with another card, so a player using a force card discards two cards in the same round. A force card allows a player to change their opponent's move. So if Player 1 throws a defect and a force card, and Player 2 also throws a defect, then Player 1 can use the force card to change the value of Player 2's card to cooperate. The force card does not physically change the cards, rather it changes the payoffs to each player. So instead of the payoff being (D,D), instead it becomes (T,S). If both players use a force card in the same round, then the payoff is (0,0). This addition of this fairly simple mechanic changes the game dramatically.

When we considered the case that $T + S = C + D, T > S, C > D$, in the DBPD, each player achieved the same score and that score did not depend on the strategy either player used. If we introduce the force card, then we will find that the strategy used by the players can matter in the final score. We can show this with an example: each player has 4 cards, 2 C's and 2 D's, with 1 force card, and the game is strict. Figure 5.16 shows an example of when the force card changes payoffs to the players.

Round	1	2	3	4
Player 1	FC	D	D	C
Player 2	D	FD	D	C
Payoffs	(C,C)	(S,T)	(D,D)	(C,C)

Figure 5.16: Given in the table are the rounds and the moves each player made. FC or FD represents a force card being played with another standard move.

Notice that Player 1's total is $2C + S + D$, whereas Player 2's total is $2C + T + D$, and since $T > S$, Player 2 is the winner of that game. Clearly, the addition of even one new card makes this a completely different game. Adding more Fluxx-like cards can change a game even further.

Fluxx-like Rock-Paper-Scissors

Fluxx-like Rock-Paper-Scissors can be created using any variety of cards, but for now we consider Rock-Paper-Scissors-Lava-Water-Rust (RPSLWT). Rock, paper, and scissors have all the usual interactions from the original RPS game, but the other three card types have the following unusual properties.

- Playing Lava defeats all Rock cards, by "melting" them. In the N-person version of RP-SLWT, the player who lays down a lava card gets one point for every rock card played by the other players, and more importantly, removes those rock cards from the game so if a player were to gain points by playing a rock card, it would now be the case they receive no points. Any player that played Paper in that round would no longer gain points for beating Rock, since they have been removed.

- The Water card removes all Paper cards, with similar effects Lava has on Rock.

- Rust removes Scissors.

We call these types of cards *negation* cards, for their obvious effects. When analyzing a game with negation cards, we use another kind of arrow to denote the negating move. Figure 5.17 demonstrates the digraph.

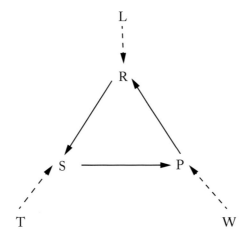

Figure 5.17: The directed graph representation of Rock-Paper-Scissors-Lava-Water-Rust. L is for Lava, W for Water, and T is for Rust. The dotted lines represent one move negating another move.

The addition of negation moves inherently changes the game, and creates a far richer strategy space to explore. In Figure 5.18, a few hands are shown for the five-player game.

Notice that in the first hand Player 3, who threw Paper as their move, should have had 2 points for both Rocks thrown, but since Player 5 threw Lava, it removed those Rocks from the payoff calculations. In the same way, Hand 3 was affected by the Rust card. In Hand 2, Water was thrown, but no Paper moves were available for it to affect, so nothing was gained by Player 5 in that instance.

Hand 1

R	R	P	S	L
0	0	0	1	2

Hand 2

S	S	R	R	W
0	0	2	2	0

Hand 3

P	S	S	W	R
0	0	0	1	2

Figure 5.18: Three examples of play with five players. The payoffs to each player, 1–5 left to right, are given underneath the cards.

5.5.5 A NOTE ON ADDING NEW MECHANICS TO MATHEMATICAL GAMES

Given here are just a few examples of how adding an extra mechanic to an otherwise simple mathematical game can drastically change the nature of that game, and increase the size of the game space, often by several orders of magnitude. Extra game mechanics need to be added sparingly and with a great deal of forethought about what they will do to the game, especially Fluxx-like mechanics. This is an area of exploration that is currently wide open, and probably needs new techniques to fully analyze what an added mechanic will do to a game. Repurposing existing techniques, such as the digraph, for analysis is a start, but this area will likely need discrete analytic and computational approaches before any real headway can be made.

5.6 EXERCISES

1. Find the Nim sequence of the subtraction game that has the subtraction set $S = \{1, 4, 5, 8\}$. If the game starts with 1,234 tokens, which player wins?

2. Find the Nim sequences of the subtraction games with sets $S_4 = \{1, 2, 4\}$, $S_5 = \{1, 2, 5\}$, $S_6 = \{1, 2, 6\}$. Is there a pattern that develops? Does it work for S_3?

3. Consider the Nim position (14,11,22,21,17). Is this a winning or losing position in Nim? Find all moves that could be made to move the game into a losing state for your opponent if it is your move. Is this a winning or losing position in Nim_4?

4. Play a game of Sprouts with five dots. Find the winning strategy for Player 1. How many moves does it take to finish the game?

5. Play a game of Sprouts with six dots. Find the winning strategy for Player 2. How many moves does it take to finish the game?

6. Consider the Graph Domination Game. Play against a friend (or enemy), and describe what kind of strategy you used. Did you win? Analyze your strategy against your opponent's, and determine why the winning strategy was superior. Pick one of the open questions and attempt to give an answer.

7. Squares: Two players are facing a square piece of paper that measures 1 m x 1 m. The first player draws a square whose center is in the middle of the larger square, such that the sides of the big and small square are parallel. The second player then draws another, larger, square such that the corners of the first square just touch the midpoints of the edges of the second square. The first player then draws a third square larger than the second, such that the corners of the second square touch the midpoints of the edges of the third square, and so on. See Figure 5.19 for a visual representation.

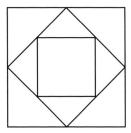

Figure 5.19: Squares: two players.

Here is the game: The person who draws the square whose edges or corners go beyond the 1 m x 1 m edge loses. If the first player is restricted by the rule that they can draw the first square with a non-zero area no larger than 16 cm^2, can the first player pick a square size that guarantees a win? What size of the initial square guarantees the second player will win?

8. Pack: Two players are facing another square piece of paper that measures 10 cm by 10 cm. The first player draws a 1 cm line segment on the piece of paper such that the every point on the line segment is at least 1 cm away from the edges of the square. The second player then draws a new line segment such that every point on the line segment is at least 1 cm away from the edges of the square, and every point of the line segment that the first player put down. And so on. Each line placed on the paper must have every point on the line segment at least 1 cm away from the edges, and every point of every other line segment already drawn. The last player who can put a line down wins. Is there a strategy that guarantees Player 1 will win? What about Player 2?

9. Find the payoffs to the five players in FDBRPS in each of the following hands:

   (a) RRPSW

   (b) RPSRS

   (c) WPRLT

   (d) TRSPS

10. Create a digraph for the following game (bonus points if you can draw it with the minimal number of crossing edges):

    • Move A beats Move C, Move E, and Move F, with payoffs of 3, 4, and 5, respectively.

    • Move B beats A and D, payoffs 2 and 7.

    • C beats B, payoff of 3.

    • D beats A and C, for 3 and 4.

    • E beats D and F, for 1 and 2.

    • F beats B and C, for 4 and 4.

    (a) Find the local balance of each move.

    (b) In our work on local balance, the payoffs are not mentioned. Come up with a way to incorporate the payoffs into the local balance of a move. Does this way change the ordering of the local balances of the moves? Does this make sense?

11. Imagine you are playing simultaneous deck based divide the dollar. You and your opponent have finite decks with identical distributions of cards: $\frac{1}{4}$ of the deck is made up of 1's, $\frac{1}{2}$ of the deck is made up of 2's, and $\frac{1}{4}$ of the deck is made up of 3's. You pull your hand of 5 cards randomly from the deck, your opponent does the same. Each deck has 60 cards total. The goal deck also has 60 cards, and has the following distribution: $\frac{1}{4}$ of the cards are 3's, $\frac{1}{2}$ of the cards are 4's, and $\frac{1}{4}$ of the cards are 5's. Come up with a winning strategy for this version of the DBDTD. Write this strategy as a finite state automata.

12. Consider an open mathematical game, or create one on your own, and turn it into a deck-based game. Define the moves and payoffs, draw a digraph, and create a card count for that game, and then write a page that justifies it as being a fun game.

CHAPTER 6

Tournaments and Their Design

6.1 INTRODUCTION

The earliest known use of the word "tournament" comes from the peace legislation by Count Baldwin III of Hainaut for the town of Valenciennes, dated to 1114. It refers to the keepers of the peace in the town leaving it "for the purpose of frequenting javelin sports, tournaments and such like." There is evidence across all cultures with a written history that people have been competing in organized competitions for at least as long as humanity has been around to write about it. This chapter offers a treatment on tournaments, a useful representation, and some new mathematics involving unique tournaments.

Note: The word *tournament* has been used in the mathematics of directed graph theory, but those series of works are not related to this topic. A tournament is a name for a complete directed graph, and while that is tangentially related, the research done in this area is not applicable to this treatment.

6.1.1 SOME TYPES OF TOURNAMENTS

Elimination Tournaments

Perhaps the most common style of tournament, this style is used in many professional sports leagues to determine playoff schedules and the yearly winner.

- Most often single or double elimination tournaments are held.

- Teams will play against one another, and are not allowed to play in the tournament if they lose once, or twice, depending on the style of tournament.

- The tournament continues until only one team remains, called the winner.

Challenge Tournaments

Also known as pyramid or ladder tournaments.

- Teams begin by being ranked by some other criteria.

- A team may challenge any team above them in the ranking. If the challenger wins, they move into that team's ranking. Every team in between the winning and losing team goes down the ladder. If the challenging team loses, the ranking does not change, or the challenging team can drop in rank.

- There are variations where teams may only challenge teams that are a specific number of ranks above them.

- The winner is the team with the top rank, after some specified number of games or time.

Multilevel Tournaments

Also known as consolation tournaments.

- Similar to single and double elimination tournaments, teams that lose one or two games are shifted down to another tier, where they play other teams that have been eliminated from the original tier.

- Many variations and tiers can accompany this style of tournament.

Round Robin Tournaments

A round robin tournament is the style that takes the longest to complete, and it, or some variation, is most often used in league play during a sports season.

- Teams play every other team in the tournament at least once.

- Rankings are determined by some cumulative score, or total number of wins with tiebreakers predetermined.

- There are many variations to the round robin style tournament, such as the double and triple split.

There are advantages and disadvantages to each style of tournament. Elimination tournaments are best employed when there is a limited number of games that can be played, due to constraints such as time. However, a bad seeding (initial arrangement of teams and games) combined with a poor showing can knock out high quality teams early. The most commonly used heuristic to avoid poor seedings is during the first round have the highest quality team face the lowest quality team, the second highest quality team face the second lowest quality team, and so on. Challenge tournaments provide an interesting method of ranking a team, but can take a long time and because challenges have to be issued, and there is no guarantee that the final rankings will be reflective of the quality of teams. Multilevel tournaments have the advantage of allowing eliminated teams to continue playing, but ultimately share the same drawbacks as elimination tournaments. Round robin tournaments take the most time to complete, but offer the best way to truly evaluate the relative quality of teams. However, a round robin tournament of 16 teams, for example, will take 120 games to complete. With 32 teams, it would take 496 games. In games that involve time and physical constraints, round robin tournaments are not practical for large numbers of teams. However, by bringing automation into tournaments, the round-robin becomes practical again, and is the best way to evaluate a team's relative quality. The rest of this chapter will focus on tournaments using the round robin style.

6.2   ROUND ROBIN SCHEDULING

The study of round robin scheduling began with the work done by Eric Gelling in his Master's thesis [21]. Due to the graph theoretic approach of Gelling, we now include some terms that will be useful in dealing with the results of his investigation.

- A factor G_i of a graph G is a spanning subgraph of G which is not totally disconnected.

- The set of graphs $G_1, G_2, ..., G_n$ is called a *decomposition* of G if and only if:
    - $\bigcup_{i=1}^{n} V(G_i) = V(G)$
    - $E(G_i) \bigcap E(G_j) = \emptyset, i \neq j$
    - $\bigcup_{i=1}^{n} E(G_i) = E(G)$.

- An *n-factor* is a regular factor of degree n. If every factor of G is an n-factor, then the decomposition is called an *n-factorization*, and G is n-factorable.

While there are many interesting results from that work, there are a few we will focus on due to their application to round robin tournaments. The following theorem helps us count the number of 1-factorizations for a complete graph with an even number of vertices.

Theorem 6.1   *The complete graph K_{2n} is 1-factorizable, and there are $t = (2n - 1)(2n - 3)...(3)(1)$ different 1-factorizations of the complete graph on 2n vertices.*

The proof can be found in Gelling's thesis, which can be accessed here: `https://dspace.library.uvic.ca:8443/handle/1828/7341`.

For the graph K_4, for example, there are $t = (4 - 1)(4 - 3)(1) = 3$ 1-factorizations, and those are pictured in Figure 6.1.

Adding labels to the vertices, such as the numbers $\{1, 2, ..., 2n\}$ for K_{2n}, complicates matters, because the ordering of the vertices creates equivalence classes of 1-factorizations. Gelling does an exhaustive search for the equivalence classes of 1-factorizations of K_6, K_8, and K_{10}. From a combinatorics and graph theory perspective, this represents an interesting challenge to categorize them all, but from a tournament design angle it is less relevant to our discussions.

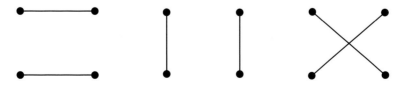

Figure 6.1: The three 1-factorizations of K_4.

Note: It is important to be specific in our discussions henceforth. A **game** will defined as one team playing against another team, defined by the 1-factorizations of the complete graph.

A **round** will be defined as the set of games defined by the edge set of a single 1-factorization. Without any physical or temporal restrictions, a round robin tournament with 2n teams in which teams play each other once requires $n(2n − 1)$ **games** to complete, and $2n − 1$ rounds.

Theorem 6.2   *In a standard round robin tournament, when $2n \leq 4$, the number of 1-factorizations is equal to the number of rounds required to complete the tournament. As $2n \rightarrow \infty$, the ratio of 1-factorizations to rounds approaches infinity.*

What we can take from this theorem is that we will be spoiled for choice in the number of possible 1-factorizations we could use to complete the 2n-1 rounds of a tournament with 2n teams. When there are no other restrictions placed on the tournament, then this is not much of a concern. When we have restrictions, like the number of available courts or fields upon which matches are to be played, then the combinatorially increasing number of 1-factorizations becomes an interesting search problem.

6.3   ROUND ROBIN SCHEDULING WITH COURTS

Designing round robin tournaments while taking courts into consideration is an interesting combinatorics and design problem all on its own. Let us use the definitions of rounds and matches defined in the previous section, and add the following:

A **court** is the physical location where a match between two teams occur.

If the tournament has 1 court, then this discussion is rather short. Every pair of teams plays on that court until the tournament is over. However, if you have more than one court, perhaps of unequal attractiveness, then the problem of choosing a set of 1-factorizations that guarantees each team will have equal playing opportunities on all of the courts becomes a difficult search problem. In fact, the graph theoretic approach, while defining the 1-factorizations as matches in a round, is no longer sufficient for this purpose.

In [22], the general methods of creating balanced tournament designs were established.

6.3.1   BALANCED TOURNAMENT DESIGNS

Let us consider a round robin tournament with $2n$ players, with n courts of unequal attractiveness. This could be due to quality of the court, location, weather, and a variety of other reasons. The tournament is to take place over $2n − 1$ rounds, and we add the condition that no team plays on any court more than twice.

- A *balanced tournament design*, $BTD(n)$, defined on a 2n set V, is an arrangement of the $\binom{2n}{2}$ distinct unordered pairs of the elements of V into an $n \times 2n − 1$ array such that:

- every element of V is contained in precisely one cell of each column; and

- no element of V is contained in more than two cells of any row.

By letting the columns correspond to player matches and the rows to court assignments, this is a representation of a round robin tournament. Optimally, each team will appear twice in each row, except for the row in which it will appear once. An example of the $BTD(3)$ is given in Figure 6.2.

5	1	4	6	2
3	2	6	1	4
6	3	1	4	5
2	4	3	5	1
4	5	2	3	6
1	6	5	2	3

Figure 6.2: An example of a $BTD(3)$.

A way to construct $BTD(3)$ is given in the following algorithm.

Create an initial round of play between the 6 teams, randomly or otherwise. Place the pairs in the first column of the lattice shown in Figure 6.2.
Call the first pairing Block 1 (B1), the second pairing Block 2 (B2), and the third block Block 3 (B3). The first team of B1 will be denoted B11, the second team of B1 denoted B12, and so on. In round 2 of the tournament, B11 plays B21, and is placed in row 3, B22 plays B32 and is placed in row 1, and B12 plays B31 and is placed in row 2.
Round 3: B11 plays B22 row 3, B21 plays B31 row 1, and B12 plays B32 row 2.
Round 4: B11 plays B31 row 2, B12 plays B22 row 3, and B21 plays B32 row 1.
Round 5: B11 plays B32 row 2, B21 plays B12 row 3, and B31 plays B22 row 3.

Without loss of generality, the steps in the previous algorithm can be rearranged in any order. There are some notes about generalization that can be mentioned here: notice that after the initial pairing, teams from Bn never appear in row n again. If the number of teams is evenly divisible by three, then we can generalize the algorithm to include a choosing method to create sub-blocks that will fill the lattice. If the number of teams does not have 3 as a divisor, then care will have to be taken in generalizing the algorithm. There are no balanced tournament designs with n being even, and this can be seen using the pigeon-hole principle.

Theorem 6.3   *There exists a BTD(n) for every odd positive integer n.*

In [22] a proof is given for this theorem and a way to construct certain cases of n, but it is heavily accented with group theory beyond the scope of this course.

6.3.2 COURT-BALANCED TOURNAMENT DESIGNS

We now move on from balanced tournament designs. Notice that the $BTD(n)$ does not guarantee a team will be able to play on their most desirable court as many times as other teams will. To counter this, we move our focus to balancing the number of times a team will play on a given court in a round robin tournament [23]. A court-balanced tournament design of n teams, even, with c courts, denoted $CTBD(n, c)$, is a round robin tournament design such that every element of V (the set of n even teams) appears the same number of times in each row.

- Consider some positive integer t, and let t be the number of columns in the new tournament construction.

- Let α be the number of times each team will appear in a given row.

The necessary conditions for the existence of a $CBTD(n, c)$ are:

1. $ct = \binom{n}{2}$,

2. $1 \leq c \leq \lfloor \frac{n}{2} \rfloor$, and

3. $c\alpha = n - 1$.

The third condition forces the court balancing property, and we'll focus on cases where the number of courts is at least two. The bounds on t and α are given as

$$n - 1 + (n \bmod 2) \leq t \leq \binom{n}{2}, \ 1 \leq \alpha \leq n - 1.$$

When $n = 2c$, it is impossible to satisfy condition 3. However, assuming $n \neq 2c$, and the 3 conditions are satisfied, we can get the following results.

The first result deals with $CBTD(2c + 1, c)$ for all positive c. *Odd balanced tournament design of side c*, denoted as $OBTD(c)$, occur when the number of teams satisfies this condition. It is fairly easy to construct the tournament schedule for c courts, and no team will appear more than twice in any row. Consider Figure 6.3.

In case the pattern is not clear on how to construct the $OBTD(c)$, consider the first row in Figure 6.3. The top of the first row has the 7 teams listed in order, and the bottom of the first row has the seven teams listed in order, but shifted so that the bottom starts with 7. These are the pairings for that court. The second row is the result of shifting the teams on the top to the left by one spot, called a permutation, and shifting the bottom row to the right by one position. The third row is created by shifting 2 positions to the left and right, respectively. This technique can be generalized to $2c + 1$ teams. List the teams along the top of the first row from 1 to $2c + 1$, and the bottom of the first row as $2c + 1, 1, ..., 2c$. Then shift the top of the first row to the left 1 position, and the bottom of the first row to the right 1 position. This generates the second row. If you shift 2 positions, you produce the third row, and so on. It only requires $c - 1$ total shifts

1	2	3	4	5	6	7
7	1	2	3	4	5	6
2	3	4	5	6	7	1
6	7	1	2	3	4	5
3	4	5	6	7	1	2
5	6	7	1	2	3	4

Figure 6.3: An odd balanced tournament design of side 3.

to produce the c rows necessary to complete the tournament. The following two theorems serve as a useful guide.

Theorem 6.4   *Let n and c be positive integers satisfying conditions 1–3 and n be odd. Then there exists a CBTD(n, c), and α is even.*

The proof is found in the paper by Mendelsohn and Rodney. An example of a CBTD(9,2) is provided in Table 6.1. Notice that teams show up four times on each court, which is the value of α in this case.

Table 6.1: Example of a $CBTD(9, 2)$

1	2	3	4	5	6	7	8	9	2	3	4	5	6	7	8	9	1
8	9	1	2	3	4	5	6	7	7	8	9	1	2	3	4	5	6
3	4	5	6	7	8	9	1	2	4	5	6	7	8	9	1	2	3
6	7	8	9	1	2	3	4	5	5	6	7	8	9	1	2	3	4

There is a pattern to this construction that is very similar to the OBTD(n). There is still a shifting of the top part of the row and the bottom part of the row to get the next row, but instead of placing it in the row directly beneath, the new set of games is placed in the same row such that conditions 1–3 are satisfied. This is a very clever construction that creates a court-balanced round robin tournament design. In fact, this theorem can be extended even further.

Theorem 6.5   *Let n and c be positive integers satisfying conditions 1–3. Then there exists a CBTD(n,c).*

The proof of this theorem in the paper depends heavily on group theory, and the interested reader can find it in the paper. Despite the power offered by these two theorems, the number of pairs (n, c) that actually satisfy these conditions are fairly minimal. For example, with 100 teams only 3, 9, 11, and 33 courts satisfy conditions 1–3.

6.4 CYCLIC PERMUTATION FRACTAL TOURNAMENT DESIGN

This section is the result of trying to plan a special kind of tournament. A group of high school students were put into teams, and those teams were to compete in an Enhanced Ultimatum Game tournament. There were no courts to play on, as it was all done electronically, but organization of the tournament was complex. Each team had to play another team as both the Proposer and the Acceptor, so there were to be two meetings between each pair of teams. Also, I wanted every team to be playing during each round (this was only possible because the teams were even), and I didn't want the teams playing their matches with changed roles back to back. I wanted to minimize the number of rounds as the last objective, since I could only count on so much attention span and time. This led to the following generalized question.

Question 1. Suppose we have n teams that are to participate in a tournament. A match is played between 2 distinct teams. During a match, each team takes on a single role, offense or defense, and cannot change that role during the match. A round consists of the matches played between designated pairs of teams. Is there a way to organize the tournament such that each team faces one another at least once, with other matches in between, and plays as offense or defense an equal number of times, while minimizing the number of rounds?

To answer this question, we begin by recalling the definition of a permutation.

Definition 6.6 A **permutation** σ, of a set of n elements X, $\sigma : X \to X$, such that σ is a bijection.

For example, given the set $X = \{1, 2, 3, 4, 5\}$, a permutation could be given by:

$$\sigma(1) = 5, \ \sigma(2) = 3, \ \sigma(3) = 2, \ \sigma(4) = 4, \ \sigma(5) = 1.$$

$\sigma(X)$ would be given as $\{5, 3, 2, 4, 1\}$.

Definition 6.7 Let X be a set of n elements and $\sigma : X \to X$. Let $x_1, x_2...x_k$ be distinct numbers from the set X. If

$$\sigma(x_1) = x_2, \ \sigma(x_2) = x_3, \ldots, \sigma(x_{k-1}) = x_k, \sigma(x_k) = x_1.$$

Then σ is called a **k-cycle**.

For a set $X = \{1, 2..., n\}$ a more compact representation of a cyclic permutation is $\sigma = (1\ 2\ 3 \ldots n\text{-}1\ n)$, indicating that the element in position 1 will now occupy position 2, the element in position 2 will now occupy position 3, and so on, and the element in position n will occupy position 1. Thus after the permutation is applied, the set will now be $\sigma(X) = \{n, 1, 2..., n-1\}$.

6.4.1   CASE 1: $n = 2^k, k \in \mathbb{Z}^+$

It is easiest to construct a round robin style tournament that satisfies the conditions of Question 1 when the number of teams is a power of two.

Theorem 6.8   *When the number of teams is a power of two, the minimum number of rounds to fulfill the requirement that each team play every other team at least once and take on the role of offense or defense an equal number of times is $2^{k+1} - 2$.*

Proof. The proof is contained within the tournament construction. First, if the teams have not been ordered, uniquely order them using positive integers, such that they form the set $T = \{1, 2, 3, ..., n\}$. Then, create two new sets, $T^{O1} = \{1, 3, ..., n-1\}$ and $T^{E1} = \{2, 4, 6, ..., n\}$, standing for *the first odd team set* and *the first even team set*, respectively. Once the first odd team set and the first even team set are determined, place the teams in the manner drawn in Figure 6.4.

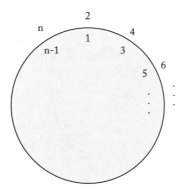

Figure 6.4: The odd numbers are placed around the inside of the edge of the circle, the even numbers are placed around the outside of the edge of the circle such that all the numbers line up.

The way the circle has been constructed designates the first round of matches. Without loss of generality, let the odd numbered teams play offense in the first round. Once the first round is complete, we "turn" the circle to the right to create the match ups of the next round, as in Figure 6.5.

In the second round, the even teams take on the role of offense. Each turn of the circle creates a new matching of teams from T^{O1} against teams from T^{E1}. The number of rounds this construction will produce is 2^{k-1}, since half of the total number of teams will play against one another, and a team cannot play against itself. Each team will have also taken on the role of offense and defense the same number of times. The circle representation is illustrative, but cumbersome, so I now switch to a representation using cyclic permutations. Let σ_{O1} be the permutation such that $\sigma_{O1} : T^{O1} \to T^{O1}$ and σ_{E1} be the permutation such that $\sigma_{E1} : T^{E1} \to T^{E1}$. Fixing T^{E1} on the line beneath $\sigma_{O1}(T^{O1})$, this gives the matches of the first round, and

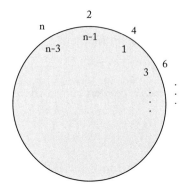

Figure 6.5: The odd numbers have all shifted as the circle has been turned, while the even numbers remain stationary.

directions on how to construct the matches in the second round, shown in Figure 6.6. Permuting T^{O1} will create the match ups indicated in Figure 6.7. While keeping T^{E1} fixed, continually permuting T^{O1} will generate the same set of matches as the circle construction.

$$
\begin{array}{cccccc}
(1 & 3 & 5 & \ldots & \text{n-3} & \text{n-1}) \\
2 & 4 & 6 & \ldots & \text{n-2} & \text{n}
\end{array}
$$

Figure 6.6: A cyclic permutation representation of the circle technique, displaying the first round of matches.

$$
\begin{array}{cccccc}
(\text{n-1} & 1 & 3 & \ldots & \text{n-5} & \text{n-3}) \\
2 & 4 & 6 & \ldots & \text{n-2} & \text{n}
\end{array}
$$

Figure 6.7: A cyclic permutation representation of the circle technique, displaying the second round of matches.

So far, this has generated 2^{k-1} rounds in which each team has played offense and defense an equal number of times. The second part of the tournament construction requires splitting the team sets T^{O1} and T^{E1} in half, forming new team sets T^{O2} and T^{O3}, T^{E2} and T^{E3}. All of these new team sets will be of size 2^{k-2}. Without loss of generality, let

$$T^{O2} = \left\{1, 3, ..., \frac{n}{2} - 1)\right\}$$

$$T^{O3} = \left\{\frac{n}{2} + 1, \frac{n}{2} + 3, ..., n - 1\right\}$$

$$T^{E2} = \left\{2, 4, ..., \frac{n}{2}\right\}$$

$$T^{E3} = \left\{\frac{n}{2} + 2, \frac{n}{2} + 4, ..., n\right\}$$

and fix T^{O2} atop T^{O3}, and fix T^{E2} atop T^{E3}. This will be the first round of matches. Then cyclically permute T^{O2} and T^{E2} in the manner described in Figure 6.6, while keeping T^{O3} and T^{E3} fixed. Let T^{O2} and T^{E3} take on the role of offense in this next set of rounds, which will total 2^{k-2}.

Assuming there are still enough teams in the team sets to evenly split, continue the splitting and cyclic permutation (or the circle, if you prefer) construction until there are only team sets of size 1 left. At this point, it should not be difficult to see that $2^k - 2$ rounds have passed. Since there is only one more round of matches left, it becomes clear that half of the teams will have taken the role offense one more time than defense after the final match has been played, giving a total of $2^k - 1$ rounds. This is the minimum number of rounds required to guaranteee a team is matched against every other team if the number of teams is 2^k, since a team cannot play itself. To correct this imbalance, retain all of the orderings of rounds and play all of the matches again, but this time reverse the offense and defense designations of all of the matches. This guarantees that each team takes on the role of offense and defense the same number of times. Since the number of rounds is doubled, $2(2^k - 1) = 2^{k+1} - 2$ is the of rounds necessary and sufficient to fulfill the condition that each team faces one another at least once, and plays as offense or defense an equal number of times, while minimizing the number of rounds.   □

CASE 2: $n \in \mathbb{Z}^+$

The case where the number of teams is not a power of 2 requires the following lemmas.

Lemma 6.9   *If a team set has two elements, it requires two rounds for each team to have played one another and have taken the role of offense and defense an equal number of times. Each team will have played two matches. If a team set has three elements, it requires six rounds for each team in that team set to have played one another and taken on the role of offense and defense an equal number of times. Each team will have played four matches.*

Proof. Consider a team set with teams labeled 1 and 2. Without loss of generality, Team 1 takes the role of offense in the first round, Team 2 takes the role of offense in the second round.

Now consider a team set with teams labeled 1, 2, and 3. Without loss of generality, Team 1 is matched with Team 2, then matched with Team 3, and finally Team 2 is matched with Team 3. Due to the definition of round, each match is also a round in this case. Team 1 will take on the role of offense twice, Teams 2 and 3 will take on the role of defense twice. Now play the same matches again, but switch the roles during the matches. Three games from the first set of matches plus three games from the second set of matches is six rounds. Each team plays the other team twice, totaling four matches per team.   □

Lemma 6.10   If $|T^A| - |T^B| = \pm 1$, where T^A and T^B are the two team sets created by splitting an existing team set, then the number of rounds required to fulfill the condition that that each team faces one another at least once, and plays as offense or defense an equal number of times, while minimizing the number of rounds, is

$$2 \cdot argmax(|T^A|, |T^B|).$$

Proof. Without loss of generality, let $|T^A| > |T^B|$. Then we can use the following construction in Figure 6.8. The team from T^A on the end of the cyclic permutation schedule will sit out each round while the teams from T^A and T^B play their matches. If $|T^B| = m$, then it will take $m + 1$ rounds to ensure that all teams from each team set are matched against one another. One round for the initial match ups, and then m iterations of the cyclic permutation on T^A. From our initial assumption, $|T^A| = m + 1$. Similar to the idea from the previous case, if we double the rounds played while switching the roles of offense and defense, then each team plays offense and defense an equal number of times. During these rounds, the teams from T^B will play two extra matches than the teams from T^A. However, those matches are made up in the next set of rounds, as shown in Lemma 6.11.   □

(A1	A2	A3	...	An-1	An)
B1	B2	B3	...	Bn	
(An	A1	A2	...	An-2	An-1)
B1	B2	B3	...	Bn	

Figure 6.8: A cyclic permutation representation where one team set has one more element than the other. Displayed are two rounds.

Lemma 6.11   Let $|T^A| - |T^B| = \pm 1$. From Lemma 6.10, the teams from T^B have played two more matches than the teams from T^A, or vice versa. After the next round is completed, the number of matches played by all of the teams in those team sets is equal.

Proof. Begin by splitting $|T^A|$ and $|T^B|$ into $|T^{A1}|$, $|T^{A2}|$, $|T^{B1}|$, and $|T^{B2}|$. Without loss of generality let $|T^A| > |T^B|$ and let $|T^{A1}| \geq |T^{A2}|$ and $|T^{B1}| \geq |T^{B2}|$. Since $|T^A| - |T^B| = 1$, T^{A1} has one more element than T^{B1}, then there is one more match that needs to be played among the A teams than the B teams. We double the number of matches to equalize the roles of offense and defense, thus creating two extra matches among the A teams than the B teams. Thus, the number of matches played over the course of two sets of rounds is equal. □

Theorem 6.12   *When the number of teams, n, is not a power of two, the minimum number of rounds to fulfill the requirement that each team play every other team at least once and take on the role of offense or defense an equal number of times is $2(n + z)$, where z is the number of rounds a team split creates team sets of uneven size. The exception is when $n = 7$ or $7 * (2k), k \in \mathbb{Z}^+$, or when a team split causes a team set size to be seven or an even multiple of seven. If this occurs, this split does not add to the z count.*

Proof. Begin by assuming the number of teams in the tournament is even but not a power of 2. Based on the construction given in the proof of Theorem 6.8, the number of rounds to play all of the necessary matches will be $\frac{n}{2}$. After the first set of rounds is complete, then a team split occurs. If the team split results in all teams being of size $\frac{n}{4}$, then the number of rounds to play all of the matches will be $\frac{n}{4}$, and so on. However, since n is not a power of 2, at some point a team split will result in the sizes of one of the new team sets being odd. Once that occurs, every split of that odd-sized team set afterward will result in team sets of unequal sizes. Once there are team sets of unequal sizes, we know from Lemma 6.10 that the number of rounds necessary to play all of the matches in this case is $m + 1$, where m is the size of the smaller team set. Thus, every time a team split creates team sets of unequal size, one more round is necessary to ensure every team plays against one another. Let z be the number of times a team split creates team sets of unequal size, and let n be the number of teams in the tournament. Then it requires at least $n - 1$ rounds to ensure every team plays against one another. n is not a power of 2, so during the last round of play there will be at least one team set of size three. From Lemma 6.9, we know that it requires an extra round for team set of three to play against one another. This extra round means that it will require n rounds for n teams to play one another. Adding the necessary rounds created from unequal splits, the sum is $n + z$. As it was necessary in the earlier construction, doubling the number of rounds while reversing the roles guarantees satisfaction of the conditions set out in Question 1, thus the number of rounds necessary is $2(n + z)$.

If we assume n is odd, then every round will result in an uneven team split, and following reasoning from the case where n is even, every time there is a team split one more round must be added to accommodate the uneven number of teams.

The exception is when a team size is seven. If we split seven, we get team sets of four and three, requiring four rounds to play all of the matches. Once we split again for the next sets of rounds, we now have team sets of size three, two, and two. The teams from the size three split require three rounds to play one another. The two team sets of size two created from the size four split require two rounds to play one another. Team sets of size three do not need to be split, and when the size two teams are split into team sets of size one, they require one round to play one another, still totalling three rounds. Thus, four rounds from the first split plus three rounds from the second and third split total seven rounds. No extra rounds are necessary for a team set of size seven. Any even multiple of seven will have this property as well, since all splittings will eventually create team sets of size seven. □

6.4.2 A RECURSIVE GENERATION OF THE MINIMUM NUMBER OF ROUNDS

There is a recursive method to generate the minimum number of rounds required to play a tournament that satisfies the conditions given in Question 1.

Theorem 6.13 *Let n be the number of teams in the tournament and let $r(n)$ be the minimum number of rounds to complete the tournament. Then*

$$r(n) = \begin{cases} n + r(\frac{n}{2}) & \text{if } n \text{ is even} \\ r(n+1) & \text{if } n \text{ is odd} \end{cases}.$$

Proof. Consider when n is even. The first split creates two team sets of size $\frac{n}{2}$. It will take $2 \cdot \frac{n}{2} = n$ rounds to play all of the matches required to satisfy the conditions laid out in Question 1. Once that is completed, assuming there are enough teams in the team sets to split again, the required number of rounds will be equivalent to a tournament of half of the original size.

Consider when n is odd. By Lemma 6.10, the number of rounds required to have the two team sets play each other is the same as if we had started with $n + 1$ teams. □

6.5 EXERCISES

1. Provide a lattice that satisfies the conditions of a *BTD*(5).

2. The number of pairs (n, c) that satisfy the conditions satisfying a *CBTD*(n, c) for n teams have been explored in sequence A277692, in the Online Encyclopedia of Integer Sequences (https://oeis.org/A277692).

   (a) Investigate the distribution of this sequence by finding its mean, mode, and median for the first 1,000 terms.

   (b) How many values of n only have 1 court that allows a *CBTD*? What do those values of *n* have in common?

(c) Using software provided or creating your own, plot the number of teams vs. the terms of the sequence. What do you notice?

(d) Perform some other analyses you think appropriate. Can you get an idea of why this sequence behaves as it does?

3. Create a tournament schedule for 22 teams using the cyclic permutation fractal tournament design. How many games does it take to complete? How many rounds does it take to complete? Which values of c satisfy a $CBTD(22, c)$?

4. Create a payoff function for a participant in the VGT. This function should include:

   - the choices: vaccinate or do not vaccinate;
   - the number of children, X;
   - the background infection rate, b;
   - the vaccination rate, v;
   - the transmission rate, t; and
   - and the number of people involved in the game, N.

5. Refer to the Vaccination Game Tournament described in Appendix A.1.2. Consider yourself as a player in the VGT. Fix the values of X, b, v, t. For what values of N do you choose to vaccinate? When is the risk acceptable to not vaccinate?

6. The VGT assumes complete connectivity throughout the population. Consider yourself as a player in the tournament. How does your strategy change if you know that there are small world network effects in play?

7. Do some research on: "At what size N does a population go from exhibiting small population effects to something more akin to the law of mass action?" Is there anything in the literature that gives a suggestion?

CHAPTER 7

Afterword

7.1 CONCLUSION AND FUTURE DIRECTIONS

This book barely scratches the surface of what can be studied through a game theoretic lens, but it was my intention to bring to the forefront some subjects of interest that represent new avenues to explore. Classical game theory extends well beyond what is mentioned in this book, and if you are more interested in that area, start with John von Neumann and Oskar Morgenstern in *The Theory of Games and Economic Behavior*. The text by Herbert Gintis, *Game Theory Evolving* is an excellent work and has dozens of examples to illustrate the deeper theorems of classical game theory.

Evolutionary game theory is an area that is not explored in this book, due to its requirements that a student must have an understanding of dynamical system theory. For the more mathematically inclined student, evolutionary game theory represents a way to combine Darwin's theory of evolution with dynamical systems to get a sense of how a population of game players will behave over time. *Evolution and the Theory of Games*, by John Maynard Smith, is the seminal work in this field, introducing the Evolutionary Stable Strategy, and has spawned countless directions of interesting research in evolutionary game theory.

If you found that you were more interested in mathematical and combinatorial games, *Winning Ways For Your Mathematical Plays*, by Elwyn R. Berlekamp, John H. Conway, and Richard K. Guy is the place to start. Another place to look for mathematical games is Martin Gardiner's *Mathematical Games*, which was a column in the Scientific American journal, has been turned into a collection now available on Amazon, `https://www.amazon.ca/Mathematical-Games-Collection-Scientific-American/dp/0883855453`. There are many journals featuring research into combinatorial game theory, and it represents a rich field to browse. If you are interested in discrete mathematics, combinatorial game theory is a great place to play. As for what I define as mathematical games, I have found no cohesive set of works currently available to read. There are some journals that are interested in the topic, such as *Games and Puzzle Design*, and I hope that more arise in the future.

The idea of the tournament has been explored in the economics literature, but as a way to incentivize employees, rather than studying the structure of tournaments. There has been some work buried fairly deeply in the combinatorics literature, and the rare paper in journals on video game design. The artifical intelligence community that is interested in teaching computers how to play games has explored different kinds of tournaments, and how their structures can affect performance and learning by a computer agent. If you are interested in combinatorics and in competition, this is a natural place to explore. Even better, this is an open topic, with many tournament structures unexplored and not yet optimized.

The techniques learned from game theory can be broadly applied. One of the more interesting and insightful applications of game theory was to the implementation of social policy regarding social programs, like minimum income. James M. Buchanan, the Nobel Prize lecturer, wrote a book in 1975 called *Altruism, morality and economic theory*. In this book he wrote about the *Samaritan's Dilemma*, and the game theoretic technique he uses leads to the conclusion that a minimum income should be provided for everyone. With the experiments in minimum income taking place in a few different countries around the world, it will be interesting to see if Buchanan's theories, while mathematically sound, will have any bearing on the outcome of those experiments.

Game theory also has an important place in health policy decisions. Having at least some understanding of how people will behave on average when it comes to virus pandemics could save governments millions of dollars, and, without exaggeration, save lives. Questions about finding the correct incentive structure in getting people to vaccinate and knowing where and with whom the vaccine will be most effective is right in the province of game theory. These techniques are being used in policy decisions around the world, and there is still much work to be done in this area.

Finally, the combination of the fields of evolutionary computation and game theory has already taken place, but is in its infancy. Whether it is classical, evolutionary, mathematical, or any other variety of game theory, evolutionary computation represents a novel way to investigate those topics. Due to its unorthodox exploratory nature, evolutionary computation is a way to go beyond the Nash equilibrium or the Evolutionary Stable Strategy and further explore aspects of games that standard techniques have may have missed. The combination also represents a way to explore the design of novel games in a way that is not offered by any current incarnation of game theory. The video game and artifical intelligence area is a natural place to explore these concepts, and while some work has already been done, there is much to do.

APPENDIX A

Example Tournaments

Offered in this appendix are some example tournaments and some thoughts on how to organize and run your own tournament in a classroom.

A.1 EXAMPLE TOURNAMENTS

Four tournaments offered in a game theory class are outlined here.

A.1.1 THE ENHANCED ULTIMATUM GAME TOURNAMENT

The Enhanced Ultimatum game represents an interesting way to both explore mathematical thinking and picking a successful strategy when you have an idea of how other people are going to play. This tournament has gone through two iterations so far. In the initial tournament play, teams would have a face to face interaction, but this was both time consuming and led to bargaining, which changed the nature of the game entirely. In the interest of removing as many confounding factors as possible, we have since moved the EUG tournament to a strictly computational format. Two versions of this tournament now exist.

Version 1

In this version, determination of rounds and matchups is necessary, in the fashion given by cyclic permutation fractal design. Each team faces some other team, taking on the role of Proposer or Acceptor. The Acceptor team delivers a demand electronically to the Proposer, the Proposer responds with a proposal. The minimum accept is predetermined by the Acceptor team, and if the proposal exceeds or equals the minimum accept the deal goes through. Each team's total score is displayed so every team knows the score of every other team, but not their individual match results. The number of rounds that has passed is also displayed.

Pros:

- Allows teams to develop a sense of strategy over the course of the tournament.

- Generates a sense of excitement among the teams doing well; often there is fairly close competition.

- Fosters a team's need to communicate with one another, and decide on a course of action based on their current score, and the score of their opponent.

- Strategies can be changed during the tournament, so a sense of the evolution of strategies can be determined.

Cons:

- Teams that are losing often lose interest can sabotage the rest of the tournament with unreasonable demands.

- Even with computational aid, running a tournament with many teams can still take a long time. With 32 teams, for example, 62 rounds have to pass for a complete tournament to take place. If each round allows 5 min total for deciding demands and proposals, we are still looking at 310 min, or just over 5 h.

- Requiring repeated user input is rarely a smooth process. This can be alleviated by using techniques such as default values, but that takes away from the whole point of the tournament.

   For the most part, version 1 only works for a smaller number of teams, and even then it can be problematic.

Version 2

Version 2 of this tournament requires that teams secretly submit an array of values, or a function, $p = f(d)$ (proposal as a function of the demand), along with a minimum accept and a demand to the tournament. These values are static, and do not change. Assuming that there are $20 on the table to split, teams submit an array of 21 values. The first 19 values are what they would propose if they received a demand of that value's position. So, for example, if the value in position 1 was 1, that would mean the team would offer 1 if the other team demanded 1. If the value of position 10 was 8, that would mean the team would offer 8 if the other team demanded 10, and so on. The last two values are the minimum accept and the demand. If teams submit a function, they must specify the function and their minimum accept and demand. Once every team has submitted their functions or their arrays, the tournament can be run using a computer. Roles, demands, and proposals are all predetermined, so the actual running of the tournament tends to go very quickly, even in Excel, and the results of the tournament are often known nearly instantly for even a large number of teams.

Pros:

- Tournaments can be run immediately. This allows for teams to discuss a strategy before submitting it, and there are no problems of loss of interest in the results.

- Instead of having one tournament, this allows for the running of several subtournaments in a very short time span. Allowing teams to see the results of the each subtournament along with allowing them to change submissions is a nice way to see evolution taking place, and with the computational aspect there is no need to wait a long time to see results.

- Encourages strong mathematical thinking about average payoffs and likelihoods of demands.

Cons:

- Once a decision is made about an array or a function, it is final. There is nothing a team can do about their strategy once it is entered into the tournament.

- No chance of evolution of strategy over the course of a single tournament.

- Teams may have different strategies they'd like to employ when facing one team or another, this version does not allow for that.

For the most part, I find version 2 of the tournament superior to version 1. If you treat the subtournaments as rounds of a larger tournament and allow teams to resubmit strategies from round to round, this is a good appoximation to evolution of strategy that can take place during a tournament similar to version 1.

A.1.2   THE VACCINATION GAME TOURNAMENT

The Vaccination Game Tournament (VGT) does not involve teams playing against specific teams, rather all teams are competing against one another every round. The VGT was created by Scott Greenhalgh and myself, and is based on the following, admittedly ludicrous, scenario.

- You are an immortal and you will have X children every generation. There is a terrible mutating disease that scours the land before your children are born.

- You get to make a choice between two options, vaccinate or not, before the epidemic hits. You know it shows up without fail.

- If you choose to vaccinate, it is 100% effective, but you lose pX children, where $0 < p < 1$. The rest are guaranteed to live through that generation. Vaccinations are only good through one generation. After your children survive that first generation, they are safe from the disease to live their lives.

- If you do not vaccinate and do not get the disease, all X of your children live through that generation. If you get the disease, all of your children die that generation. The disease goes away before the next time the epidemic returns.

- The immortal with the most children after some number of generations is the winner of the tournament.

- Everyone makes their decision about whether or not to vaccinate before the disease runs rampant.

- If you vaccinate, you cannot be infected with the disease, and cannot infect someone else.

- You can be infected by the background initially at some rate b, if you don't vaccinate.

- Then everyone interacts with everyone, and has a chance to pass on the disease if they were infected to begin with, with transmission rate t.

- At the end of the interactions, those who are not infected get their children, minus pX if they vaccinated.

- Once everything is sorted, a report is given on a percentage of the population: how many people vaccinated and how many people got sick of that subpopulation. This is to simulate the fact that imperfect information is usually available

This tournament is interesting, because while the rounds are independent, meaning being sick last round does not affect this round, the distribution of information can affect the strategy of a given player round to round. For example, imagine a player chooses not to vaccinate, and does not get sick in a given round. The report is issued after the round is complete, and she notices that only a small percentage of people vaccinated, and many did get sick. She may decide that she got very lucky that round and vaccinate the next one. In the same vein, if she vaccinates, and then notices that everyone is vaccinating, she may choose to ride the herd immunity effect in the next round and not vaccinate. The algorithm of how the VGT runs is given below.

Collect player choices about whether to vaccinate or not
Check to see who is infected by the background infection
Have players interact, determine who becomes infected by contact if they weren't already infected by the background
Determine payoffs based on outlined criteria
Give players a chance to make a new choice
Repeat

The algorithm is simple, along with the game, but it can produce some interesting and complex behavior from the population that plays it.

This tournament drives theoretical biologists up the wall, because often it does not conform to what the mathematics says should happen: just about everyone who does not vaccinate should get sick. The law of mass action, usually applied to chemistry, has been used in the biological literature for many years modeling the spread of pathogens throughout a population. However, there must be sufficient *mass* and connectivity for the law to work. In the VGT, if there aren't enough people to satisfy that requirement, the stochasticity of the game makes it very difficult to predict what will actually happen, even when how many people are playing, infection rates, and transmission rates are known. This is another aspect of a recent phenomenon of evolutionary computation called *small population effects* [24]. It captures the idea that small populations are prone to unusual results when placed in competitions, like tournaments or evolution simulations. An open question is the following:

At what size N does a population go from exhibiting small population effects to something more akin to the law of mass action?

A.1.3   A DIFFERENT KIND OF IPD TOURNAMENT

The main differences between this tournament and Axelrod's Tournament and its variations run by other people is that it includes a spatial structure and it involves human input. There has been some work done on adding a spatial structure to the study of the Iterated Prisoner's Dilemma. Those investigations were all done from an evolutionary dynamics perspective, rather a tournament perspective. In Axelrod's original formulation, every strategy played against every other strategy a specified number of times, for a specified number of rounds. In the second tournament he changed the number of rounds by having a certain probability that a game could end on any given round. However, the tournament still maintained that every strategy would face off against every other strategy. A spatial structure defines the connectivity between agents in a game, and restricts which agents can play against one another. There are several representations possible, but a combinatorial graph is a very useful one. As of the time of this writing, there has been a good deal of research on imposing a spatial structure on an evolving population of agents that play the iterated prisoner's dilemma, [3, 4, 5] to name just a few. However, there have been no attempts to incorporate structure into an IPD tournament played by agents that have occasional human input. The tournament is conducted as follows.

1. Originally, I had used the following schematic to build game boards: if there are N teams, each team starts with N agents, all of whom will employ the same strategy. Those agents will be randomly placed on an N by N lattice structure, called The Board, such that the opppsite ends are considered connected, forming a torus. If N is small, you may want to consider increasing the number of agents allotted per team. My personal recommendation is a minimum N of 25. *This was a mistake.* Having that many agents available meant the tournament would go on far too long, and many of the participants would lose interest. Shrink the number of agents available to each team to a small number, perhaps 5. This causes the board to no longer be an N by N lattice, however, a rectangular board approaching a square is still possible.

2. Each team will select a single location on the Board in which to play, and will play the IPD against their four neighbors to the north, east, west, and south, called the von Neumann neighborhood. The agent may face opponent agents from other teams, or its opponent may be an agent from its own team. The number of rounds should be at least somewhat randomized. I personally use 150 rounds ± 5. It may also be interesting to use the convention created in Axelrod's second tournament, where the game has a chance of ending after each round.

3. The scores are tallied for each team when the four games are completed for each agent that was selected to play, and then sorted. Have ties broken at random. The top third of teams

gets to replace another team's agent in the neighborhood in which they just played with a new agent from their own team. So if there are 30 teams in the tournament, after one round the top 10 teams now have $N + 1$ agents on the board, and the 20 poorer performing teams now have $N - 1$ teams on the board.

4. Repeat steps 1–2 as many times as desired.

5. A team's score in the tournament is how many agents they currently have on the board, **not** their score from round to round. The team with the most agents on the board after a pre-specified time is the winner.

At certain points during the tournament, such as the end of every class week, allow teams to update their strategy. There is also the question of allowing teams to see one another's strategy during the week. I allow it. Copying successful behavior has been done, and is currently being done, in just about every human endeavour. It also creates interesting tournament dynamics. As teams begin to converge on the most successful strategy, it behooves teams to stay innovative in order to maintain their edge. It is this aspect of the tournament, more than anything, that forces the students to apply their knowledge in creative ways. From the reports of my students, it is also makes it more fun.

A.2 SOME THINGS TO CONSIDER BEFORE RUNNING A TOURNAMENT IN A CLASSROOM

While teaching a game theory class I had the idea that a tournament is an excellent way to encourage the learning of class material, and to investigate some new tournament properties and dynamics. Running these tournaments resulted in some interesting lessons learned, both about the tournaments themselves and how to run a tournament. Some things to consider before attempting to run a tournament:

- Keep language simple, direct, and do not assume anyone knows what you are trying to tell them. Ever.

- The tournament does not need to be short, but every team playing must feel like they have a chance at winning. Do not let it be the case that teams that have no chance of winning will stay in the tournament. Eliminate them. Otherwise, at best, teams will stop participating, and at worst they will do their best to upset your tournament.

- Automate everything that should be automated. Relying on teams or yourself to repeat tasks that would be better and more quickly done by a computer is a mistake that will cause your tournament to stagnate.

- Decide what program(s) you are going to use to help you run the tournament. I use EX-CEL, as it has the built in connectivity I desired and can handle some fairly complex agents in the form of programs. I also use Oracle, which is a database management system that allows for user access from a variety of platforms, and has a central data storage system. This is key for any kind of tournament that involves submission by teams.

- Have your teams submit strategies as either a program (FSA in some cases) or strategies that can be written as a program.

- Encourage your teams to be larger than one person. Individuals participating is fine, but a lot of excellent learning opportunities occur when trying to make a strategy clear to another teammate, or deciding what your agent's next move should be.

- Make available as much information of how the tournament is progressing as possible. This includes standings, game state, and possibly strategies. This can be done using a variety of free programs, or through a class website.

- Do not run a tournament over the class time of an entire week. It creates chaos in the course and does not allow the students enough time to truly grapple with the material.

- With computational power at its current level and cloud data storage so cheaply available, let a tournament run over the course of a semester. Dedicate a few minutes of class time every class, or have a designated time a few days a week, to further the tournament's progress. The students will have time to think about their strategies and develop better understanding of what is going on.

- Maybe most importantly, have a prize for those who do well in the tournament. Often, competitive drives in teams are crushed by laziness or other factors, so be sure to incentivize them to do their best.

Bibliography

[1] R. Axelrod, Effective choice in the prisoner's dilemma, *Journal of Conflict Resolution*, vol. 24, no. 1, pp. 3–25, 1980. DOI: 10.1177/002200278002400101.

[2] R. Axelrod, More effective choice in the prisoner's dilemma, *Journal of Conflict Resolution*, vol. 24, no. 3, pp. 379–403, 1980. DOI: 10.1177/002200278002400301.

[3] M. A. Nowak and R. M. May, Evolutionary games and spatial chaos, *Nature*, vol. 359, no. 6398, pp. 826–829, 1992. DOI: 10.1038/359826a0.

[4] M. A. Nowak and R. M. May, Evolutionary prisoner's dilemma game on a square lattice, *Physical Review E*, vol. 58, no. 1, p. 69, 1998. DOI: 10.1103/PhysRevE.58.69.

[5] K. Brauchli, T. Killingback, and M. Doebeli, Evolution of cooperation in spatially structured populations, *Journal of Theoretical Biology*, vol. 200, no. 4, pp. 405–417, 1999. DOI: 10.1006/jtbi.1999.1000.

[6] M. Sipser, *Introduction to the Theory of Computation*, vol. 2, 2006. DOI: 10.1145/230514.571645.

[7] D. Boeringer and D. Werner, Particle swarm optimization versus genetic algorithms for phased array synthesis, *IEEE Transactions on Antennas Propagation*, vol. 52, no. 3, pp. 771–779, 2004. DOI: 10.1109/tap.2004.825102.

[8] J. Von Neumann and O. Morgenstern, *Theory of Games and Economic Behavior*, Princeton University Press, 2007.

[9] C. Blair Jr., Passing of a great mind, *Life*, vol. 25, p. 96, 1957.

[10] D. Ashlock, *Evolutionary Computation for Modeling and Optimization*, Springer Science and Business Media, 2006. DOI: 10.1007/0-387-31909-3.

[11] H. Gintis, *Game Theory Evolving*, 2nd ed., Princeton University Press, 2009.

[12] G. Heal and H. Kunreuther, The vaccination game, *Risk Management and Decision Processes Center Working Paper*, no. 5–10, 2005.

[13] S. M. Ross, Goofspiel—the game of pure strategy, *DTIC Document*, 1970. DOI: 10.1017/s0021900200035725.

[14] G. Kendall, M. Dror, and A. Rapoport, Elicitation of strategies in four variants of a round-robin tournament: The case of Goofspiel, *IEEE*, 2014. DOI: 10.1109/tci-aig.2014.2377250.

[15] N. Fox, On aperiodic subtraction games with bounded nim sequence, *arXiv preprint arXiv:1407.2823*, 2014.

[16] D. Applegate, G. Jacobson, and D. Sleator, Computer analysis of sprouts, *Citeseer*, 1991.

[17] D. Ashlock and E. Knowles, Deck-based prisoner's dilemma, *IEEE Conference on Computational Intelligence and Games (CIG)*, pp. 17–24, 2012. DOI: 10.1109/cig.2012.6374133.

[18] D. Ashlock and J. Schonfeld, Deriving card games from mathematical games, *Game and Puzzle Design*, no. 1, vol. 2, pp. 55–63.

[19] J. F. Nash Jr., The bargaining problem, *Econometrica: Journal of the Econometric Society*, pp. 155–162, 1950. DOI: 10.2307/1907266.

[20] D. Ashlock, Graph theory in game and puzzle design, *Game and Puzzle Design*, no. 2, vol. 1, pp. 62–70, 2016.

[21] E. N. Gelling, *On 1-Factorizations of the Complete Graph and the Relationship to Round Robin Schedules*, Master's thesis, 1973.

[22] P. J. Schellenberg, G. H. J. van Rees, and S. A. Vanstone, The existence of balanced tournament designs, *Ars Combinatoria*, vol. 3, pp. 303–318, 1977.

[23] E. Mendelsohn and P. Rodney, The existence of court balanced tournament designs, *Discrete Mathematics*, vol. 133, no. 1, pp. 207–216, 1994. DOI: 10.1016/0012-365x(94)90027-2.

[24] D. A. Ashlock, K. M. Bryden, and S. Corns, Small population effects and hybridization, *IEEE Congress on Evolutionary Computation (IEEE World Congress on Computational Intelligence)*, pp. 2637–2643, 2008. DOI: 10.1109/cec.2008.4631152.

[25] E. A. Stanley, D. A. Ashlock, L. Tesfatsion, and M. Smucker, Preferential partner selection in an evolutionary study of prisoner's dilemma, *Biosystems*, vol. 37, pp. 99–125, 1996.

[26] C. T. Bauch and D. J. D. Earn, Vaccination and the theory of games, *Proc. of the National Academy of Sciences of the United States of America*, vol. 101, no. 36, pp. 13391–13394, 2004. DOI: 10.1073/pnas.0403823101.

Author's Biography

ANDREW MCEACHERN

Andrew McEachern obtained his Ph.D. in Applied Mathematics in 2013 from the University of Guelph, in Ontario, Canada. He is a Coleman postdoctoral Fellow at Queen's University, and his research interests include evolutionary computation, applied game theory, mathematics education initiatives, and mathematics curriculum development for elementary and high school students.

Printed in the United States
by Baker & Taylor Publisher Services